T0305625

EmDrive

EmDrive provides a comprehensive description of the theoretical background of this emerging technology. It includes the derivation of the essential equations, provides full thruster design details, and describes the manufacture and methods of testing that would enable the work to be correctly reproduced in the appropriate research facilities.

Electromagnetic drive is a proposed method of propulsion that does not require a propellant, although it still requires fuel. It holds the potential to revolutionize renewable energy production, clean and quiet transport, and global climate control. Having evolved from numerous individual and organizational contributions, this book explains the origin and subsequent development of this theory from the original UK government requirement. The sequence of experimental devices is covered in detail, and the subsequent test results are discussed. Similar programmes in the USA and China are introduced, and the implications of recent disclosures are considered.

This book will interest industry professionals working on electromagnetic theory and experimental physics in the fields of aerospace and energy engineering.

Roger Shawyer is a British Chartered Electrical Engineer with 50 years of experience in the Space and Defence industries. He is the inventor of the EmDrive propulsion technology, and for the last 21 years, he has been the director of a small R&D company, Satellite Propulsion Research Ltd. (SPR). In 1981, Roger moved into the Space Industry, taking up a position as a Principal Engineer with Marconi Space and Defence Systems Ltd. He gained extensive project management experience as the Payload Project Manager on four Eutelsat, high-power TV Broadcast Satellites. Another major programme was the Galileo European navigation satellite system, for which Roger was Technical Manager, responsible for the early design and development of the signal structure and payload equipment. In 2001, Roger left to set up SPR Ltd. and carry out initial EmDrive Research with 5 years of UK government funding. In 2008, Roger was elected a Fellow of the Royal Aeronautical Society.

EmDrive
Advances in Spacecraft Thrusters
and Propulsion Systems

Roger Shawyer

CRC Press
Taylor & Francis Group
Boca Raton London New York

CRC Press is an imprint of the
Taylor & Francis Group, an **informa** business

Designed cover image: Jung Dong Whi, sponsored by Luna Aircraft and produced at Globe Point, Korea

First edition published 2024
by CRC Press
2385 NW Executive Center Drive, Suite 320, Boca Raton FL 33431

and by CRC Press
4 Park Square, Milton Park, Abingdon, Oxon, OX14 4RN

CRC Press is an imprint of Taylor & Francis Group, LLC

© 2024 Roger Shawyer

ISBN: 978-1-032-59900-7 (hbk)
ISBN: 978-1-032-59901-4 (pbk)
ISBN: 978-1-003-45675-9 (ebk)

DOI: 10.1201/9781003456759

Typeset in Times
by codeMantra

This book is dedicated to Margaret, whose unwavering love and support made the EmDrive story possible

Contents

Preface

This book was originally meant to be a straightforward textbook, describing in detail the emerging technology called EmDrive. The name stands for electromagnetic drive, which is a method of propulsion that does not require a propellant. It is not magic and does not require exotic new physics to explain the operation. However, the implications are extraordinary, and it became clear that simply writing a text book, based only on a number of my previous technical reports and scientific papers, would seem arrogant. As with many complex concepts, EmDrive has evolved over a long period, and that evolution has included contributions from a large number of people and organisations. Although some of the contributors cannot be identified because of commercial and national security constraints, it became clear that a full understanding of the technology would require the complete story to be told.

Therefore, this book covers the story from the original UK government requirement and the concept that resulted through a number of iterations of the theory. EmDrive has also given rise to a number of alternative explanations for the measured results. These have included theories that have invoked quantum plasma, quantised inertia, Mach effects, gravitomagnetism, and negative mass. Although these theories are interesting, they are not covered here, as each would require a separate book.

During the early research period, some of my career experiences convinced me of the real need for such technology, so I started to work on EmDrive full time. The resulting sequence of experimental devices is described in detail, and the test results are discussed. Similar programmes in the USA and China are introduced, and the implications of recent disclosures are considered. The development process in the UK produced four generations of technology, and some future applications of each generation are described. These applications, taken together, enable many of the world's present problems of renewable energy production, clean, quiet transport, and global climate control to be solved. The space vehicles described will completely change launch to Earth orbit as well as make missions to the Moon, Mars and beyond much easier and quicker. Hopefully, this book will also inspire a new generation of engineers to apply this revolutionary technology to build a more sustainable and secure world.

AUTHOR'S NOTE

This book provides sufficient information to design, build and test an EmDrive thruster. Although this is a difficult task, anyone is welcome to take up the challenge; indeed, the author's patents have now been allowed to expire to encourage wider development and exploitation. However, a high-power resonating microwave cavity is an inherently dangerous device. The experimental work described in this book was carried out with meticulous safety checks in place. Test procedures were carefully reviewed, and the equipment was subjected to a safety inspection before power was applied. During the tests, microwave leakage levels were continuously monitored, and personnel were kept at a safe distance. The author accepts no liability for any injury or damage that may occur during attempts to replicate the experimental work described.

1 The Beginning

Whenever a group of engineers gather together, usually in a pub, it is remarkable how often tales of seemingly impossible requests by customers or management are told. One such request was made in early 1975 at the defence contractor Sperry Gyroscope Limited of Bracknell, Berkshire, UK. I was working at Sperry as a senior R&D engineer, specialising in electromagnetic sensors for surveillance and weapon systems. A small group of scientists and engineers from several disciplines was put together within the advanced systems group to study a very unusual requirement for a "reactionless" propulsion system. This was initially met with the usual scepticism of people schooled in the rigours of applying physics to the real world. We were all young but well versed in the laws of conservation of momentum and conservation of energy. However, when a very senior manager told us to "think the unthinkable", we guessed there was a serious problem somewhere.

There certainly was, and it was being raised at the highest level of government. Cabinet papers, originally classified as Top Secret, were finally released in December 2020 [1]. These revealed that the much-vaunted independent UK nuclear deterrent, the Polaris missile system, was really not up to the job. The probability of a warhead penetrating the anti-ballistic missile screen around Moscow was unacceptably low. In addition, the replacement warhead programme, codenamed "Chevaline", was being delayed by problems in the propulsion system. Rumours circulated about explosions during testing and accidents occurring with the highly toxic liquid propellants. The cabinet papers revealed that the Royal Navy was very reluctant to carry such technology on their nuclear submarines. Consideration was being given to a request for American help through the transfer of their new Trident solid fuel propulsion technology. However, there was concern that such a transfer, if it could be agreed upon at all, would involve costly political trade-offs and might jeopardise the ongoing Strategic Arms Limitation talks between the USA and Russia.

At this point, someone in the Ministry of Defence (MoD) remembered a recent lecture given by Professor Eric Laithwaite. This was one of the annual Royal Institution lectures and was broadcast over the 1974 Christmas holiday period by the BBC. A remarkable demonstration was given of gyroscopic force apparently generating lift. Professor Laithwaite then speculated that this could eventually be used in a propellant-less system, which would revolutionise transport. The proposal immediately led to discussions as to whether a gyroscope was an open or closed system. It is difficult to argue that a purely mechanical device, an apparently closed system, can act as an open system and cause acceleration without acting on the world around it. Huge controversy was generated in the academic world, where unnecessary scorn was heaped upon the programme. Professor Laithwaite, of course, was on the receiving end of most of the criticism, and his reputation unfortunately suffered as a result. Having been a lucky student who had been enthralled by his fluent description of electrical machine theory during his visiting lectures at Portsmouth Polytechnic,

DOI: 10.1201/9781003456759-1

I was appalled at this treatment. However, decades later, after a cover article in the New Scientist finally broke the news about EmDrive, I understood a small fraction of the bewilderment that he must have felt. The response of people who feel that their understanding of the world is being challenged is not always kind.

Because of the political sensitivity of our study, we were not told the real reason for our work, and indeed, the released papers only hint at its existence. A draft minute from the Permanent Under-Secretary to the Secretary of State includes the following statement:

> *"Before advising you further we wish to await the results of a study which we have put in hand for completion as urgently as possible on the feasibility and implications of a solution based on a British technology and not involving the use of liquid fuel."*

Clearly, the starting point for our project was the intriguing operation of a gyroscope, which, as the company name implied, was our area of expertise. The ability of a gyroscope to establish an unmoving reference plane within a highly manoeuvrable vehicle is the essence of missile guidance. One of the areas of company research was the laser gyroscope, where no moving parts were required and which was described as a strap-down device. This was pretty controversial at the time, with some of the same criticisms that were levelled at Professor Laithwaite being used to describe laser gyroscopes as impossible. If those critics could have seen the many successful applications of laser gyroscopes in use today, they would have kept quiet. The reason why a laser gyroscope is an apparent open system with information crossing the system boundary while physically being a closed optical circuit is the fact that once launched, the laser beam travels at a velocity that is independent of the velocity of the circuit itself. The constancy of the speed of light is at the heart of Einstein's theory of special relativity and is crucial to laser gyroscopes and EmDrive.

At the time, along with my colleagues on the project, the application of these concepts to the problem at hand caused much thought. Two areas of my work came into play. Firstly, I had been developing a small, low-power radar for use in a smart anti-tank weapon. The fear of thousands of Russian tanks rolling across the North German countryside was very real in those days. We had, therefore, a well-funded, ongoing development project managed by the MoD at Fort Halstead. The work covered the full spectrum, from very low-frequency seismic sensors through acoustic, microwave and millimetre-wave systems up to passive and active infrared devices. This early work eventually led to the ARGES off-route anti-tank mine for the British, French and German armed forces, together with the M93 Hornet wide-area munition for the USA. For our team at Sperry, it meant a thorough grounding in wave theory, which was to be useful to me throughout my subsequent work on EmDrive (Figure 1.1).

Second, I completed the redesign of the Sea Dart missile accelerometer to incorporate ferrite components in the magnetic circuit. This update was applied to the magnificently titled Trunnion Tilt and Angle of Sight Unit, which became an integral part of the Chieftain Tank fire control system. As a defence contractor, one quickly became used to working on both weapons and their countermeasures (Figure 1.2).

FIGURE 1.1 Sperry experimental sensor equipment.

FIGURE 1.2 Sperry Trunnion Tilt and Angle of Sight Unit.

The interesting thing about ferrites, as with many dielectric materials, is that if an electromagnetic wave is propagated through them, the velocity is dramatically reduced. Also, the dielectric properties of the ferrite cause the wavelength to decrease, and care must be taken that any ferrite component does not become a resonant element if operated at high frequency. Although this was not a problem for the Trunnion Tilt and Angle of Sight Unit, it did offer a potential method of producing a device with different propagation velocities. The force produced by a reflected electromagnetic wave, the so-called "radiation pressure", is directly proportional to the propagation velocity. Thus, a device that incorporated two different propagation velocities would also produce two different forces. Therefore, the question arose: could this force difference produce a net force that could be used for propulsion?

The immediate response was that any net force must obey the principle of conservation of momentum. It is well accepted that a travelling electromagnetic wave carries momentum and therefore any difference in momentum, due to the difference in propagation velocities must be balanced. Could the momentum of the device being accelerated by the force difference, i.e., the thrust, provide momentum in the opposite direction and thus satisfy the conservation of momentum? This proposition caused considerable discussion, and it has a profound effect on the operation and testing of EmDrive thrusters. A major difficulty arises when the idea that EmDrive is a reactionless thruster is put forward. This is not the case. A reaction force is produced if the thruster is not allowed to accelerate, and it can be measured separately from the thrust. Indeed, if an EmDrive thruster is simply placed on a set of scales with no movement, no change in weight will be measured because the thrust is balanced out by the reaction force. This is a clear demonstration that the law of conservation of momentum is being preserved and that EmDrive is not a reactionless device. The correct term to use is that EmDrive is a propellant-less device.

Clearly, the arm-waving had to be backed up by some accepted physics and maths, and I looked for previous work in this field. In those pre-internet days, the search for relevant papers was time-consuming and involved both the company library and the British Library at Boston Spa. However, in February 1975, I struck gold with a 1952 paper by Professor Alex Cullen of University College London [2]. The paper covered the theoretical and experimental work that he had carried out on absolute power measurement at microwave frequencies. Accurate power measurement of a microwave beam is notoriously difficult, usually involving an antenna and multiple diode detectors. Professor Cullen had gone back to the basics by measuring power by detecting the force produced by a beam of microwave radiation on a flat surface. The calculation of the expected force meant deriving an equation from first principles. It involved the power and the guide wavelength of the beam when travelling through a waveguide. Interestingly, the elementary theory section of the paper derives the Maxwell equation for the force due to radiation pressure from the basic force equation used in all electrical machines and originally derived by Thompson.

This harmonisation of physics and engineering, illustrated by the paper, provided an example of the method used in the subsequent development of EmDrive. Each step needed to be based on sound classical physics theory before an experiment was carried out using devices designed using well-established engineering principles.

Remarkably, I discovered many years later a paper by one of Cullen's students, Dr. R. A. Bailey, who was working at the Royal Radar Establishment [3]. His work improved Cullen's original apparatus by multiplying the force by using a resonant cavity. This had also been a critical step in the initial concept of EmDrive. However, it was one that was to bring into play the second fundamental law of physics, that of the conservation of energy. Again, this was to prove highly significant during the subsequent operation and testing of EmDrive thrusters.

Although the use of a resonant cavity increases the forces generated, it introduces a time constant into the process. This is the time required to build up to the maximum force due to the large number of reflections produced by the in-phase travelling wavefronts. The wavefront is initially propagated into the cavity and then bounces backwards and forwards along the central axis of the cavity until its energy is completely given up by the reflector losses at each end plate. The continuous supply of wavefronts at the input leads to energy being stored within the cavity. This is not unlike the stored energy in a flywheel or, indeed, a gyroscope. The amount of stored energy in a microwave cavity compared to the initial energy of one cycle is the so-called Q factor. In cavities with superconducting end plates and walls, this Q factor can reach values of hundreds of millions, leading to high stored energy and large reflection forces. The potential for a high-thrust propulsion system was clear, but it would require the low temperatures of a superconducting cavity. Clearly, if a net force can be generated by operating a cavity with different guide velocities at each end plate, the stored energy can be converted into kinetic energy by allowing the cavity to accelerate. However, the fraction of stored energy being converted to kinetic energy must not be allowed to become too great, as this results in a drop in net force. As with all things in engineering, a design must be reached, which is a compromise to produce the optimum performance.

With the fundamental elements of a possible propellant-less propulsion concept in place, backed up by some early calculations, the study was abruptly stopped as it seemed that the Royal Navy had agreed that the Chevaline liquid-fuelled propulsion system could be made safe enough to be carried in their submarines. The Secretary of State for Defence announced the decision to proceed with a liquid-fuelled Chevaline warhead and discard possible alternatives on 18 September 1975.

In November 1977, I made a visit to Sperry HQ in New York and their research site in Clearwater, Florida, for discussions on microwave and millimetre wave technology. Of particular relevance was the work that was being carried out in both the UK and the USA on Gunn diode oscillators using resonant cavities. In October 1976, Professor Roger Jennison of Kent University had published the first of a series of papers on aspects of electromagnetic inertia [4]. The paper contained details of experimental equipment using a Gunn diode source and an open resonant cavity formed by a horn antenna and a vibrating concave reflector. The apparatus formed a floating cavity with the end plate separation controlled by stepper motors, driven by a phase-locked loop. The system was not unlike an open version of what would become a typical EmDrive cavity. However, it was used to experimentally investigate Jennison's theory of inertia. Jennison had been a colleague of W.A.S. Murray, the Head of Physics at Fort Halstead, when working at Jodrell Bank. The Kent University work carried on through 1989 (Figure 1.3).

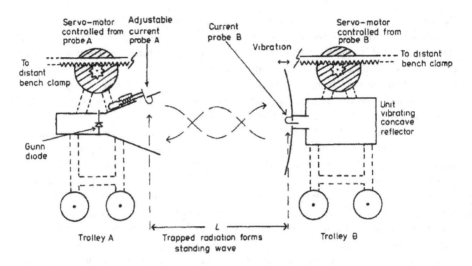

FIGURE 1.3   Roger Jennison's experimental equipment.

Whilst interest in alternative propulsion concepts continued in the academic world, my career in the defence industry continued. Highlights included field trials with the British Army, where I got to play with big boys' toys like guns, tanks and helicopters. This was great fun and included a particularly memorable test flight in a Gazelle helicopter. I was acting as the test engineer, looking after the various sensors and recording equipment on board. The flight plan included ninety-degree turns at 100 mph and 50 feet of altitude. Although this was strictly outside the normal flight envelope due to entering the dreaded "rotor flutter" condition, the pilot's attitude was refreshingly inquisitive. Being a cavalry officer, his attitude towards his helicopter was more like that towards a horse than a machine. It was a case of let's see how much seismic signal we could put into the ground. Looking at the results afterwards, it was clear that the easiest target to detect on a battlefield would be a helicopter. To survive in such a combative environment, it was clear that a new quiet propulsion system would be required, one with no discernible signature. Once again, my interest in EmDrive was encouraged. The other result of that flight was that our MoD observer, sitting in the rear seat, was violently sick. Concentrating on my instruments was apparently the only reason why I was fortunately spared the same embarrassment (Figures 1.4 and 1.5).

As well as the fun, however, there were some alarming incidents. On one occasion, I accidentally came under live artillery fire. For some upcoming trials, to measure the signatures of various armoured vehicles, a piece of particularly bumpy ground was required. The army trials team we were working with knew exactly where such a simulated battlefield could be found. So we went out to take a look at this piece of tortured English countryside. However, due to the classified nature of our work, where everything was on a strictly need-to-know basis, there had been a breakdown in communications. Together with a major from the Irish Rangers, I found myself flat on my face on the Larkhill ranges, yards away from the impact point, as the rounds came in. Many years later, at a conference reception in Toronto, I met the artillery officer who

FIGURE 1.4 Field trials with a Chieftain tank.

FIGURE 1.5 Signature trials with a Gazelle helicopter.

had been in charge of guns at Larkhill that day. Apparently, if the major had been killed, that would have been accepted as a normal training accident. However, if I had been a casualty as a civilian, there would have been an official enquiry. I suppose this was meant to be some comfort.

These rules to protect civilians were again tested when, in 1980, I was working for a company called Racal SSA in the middle of a desert, trying to establish whether some microwave sensors would operate in a more extreme environment. Suddenly, over the radio, we were informed that a war had broken out and we were to return to base immediately. The Iran–Iraq war had started, and nobody had expected it. When we got back to Amman, Jordan, there was absolute chaos. The Iraqi air force had retreated to Amman International Airport, where their fighter jets were lined up on the runway, and Iran was threatening to bomb them. As a close ally, any attack on Jordan would have dragged the UK into the conflict. My colleague and minder, a UK Intelligence officer, was issued with a sidearm for personal protection. This was entirely consistent with his position as an ex-Navy commander and all-round James Bond character. Being merely a civilian contractor, I was just given a couple of business cards from the head of the Jordanian Security Service. I was assured that these would get me out of any difficulty. On the evening before my extraction from this increasingly worrying situation, we met an interesting character who had just arrived in town. With true conspiratorial flair, in the hotel bar, he gave us his information from the battlefront. The next day, as I was going through Amman airport to catch the first refugee flight out, I was desperately hoping the business cards would work if I was stopped. I was carrying a sealed film canister in my briefcase with strict instructions not to allow it out of my hands until I got to the UK.

It was after that particular incident that I decided to look for a rather safer career path. At the time, I was also working on a number of proposals, including a signals intelligence project in Syria, a security system for a military establishment in Saudi Arabia, and an airfield perimeter security requirement in Nigeria. A long-term deployment to any of these countries was not a pleasing prospect, and so the offer of a chance to join the space industry in the UK was eagerly pursued. I was particularly fortunate to have designed a small millimetre-wave radar while at Sperry. This meant I had the right experience to work on the new EHF receiver for the proposed Skynet 4 satellite, which was my passport to an offer of a job with Marconi Space and Defence Systems at Portsmouth. It also meant that I had an appreciation for the difficulties of working with the high-precision devices that were used at these frequencies. This precision would play an important role in the very high-Q cavities that I would work with during the development of EmDrive.

It was not long after joining Marconi that the possible application of the EmDrive concept to propelling spacecraft once they were in orbit was brought into focus. I attended a meeting in which anti-satellite concerns were being expressed by a very senior military man. I was enthusiastically explaining the details of a very sophisticated anti-jamming system for the Skynet 4 communications satellite when I was cut short by a curt comment that anti-jamming techniques were all very well, but what were we going to do about *"the fellow with a few ounces of high explosive and a bag of nails"*. This was, in fact, a reference to the much-feared technique of bringing a tiny sacrificial satellite close to the target satellite and then exploding it. Only a very small explosion with a bit of shrapnel would be required to render a typical communications satellite inoperable. The classic military tactics of evasive manoeuvring would always favour the attacking satellite, with the conventional thruster signatures of the target satellite being easy to detect and track. An EmDrive thruster system, allowing continuous manoeuvring throughout the operational life of the satellite and without any exhaust signature, would clearly offer protection in this threat scenario.

With new-found enthusiasm, I resurrected the original EmDrive ideas, and practical thruster concepts began to emerge. In true inventors' tradition, early experiments were conducted in my garage, and gradually an experimental thruster hung from a converted chemical balance began to provide tentative thrust measurements. The microwave power sources were oven magnetrons, and by the time the fourth unit had expired, the technique necessary to power up the cavity without causing alarming sparks and bangs was finally established. In November 1988, the first of the EmDrive patents were filed, based on a cylindrical resonant cavity with a shaped internal dielectric. The use of a dielectric was an update on the ferrites in the original concepts. The patented device gave thrust due to the different EM propagation velocities, and hence different radiation pressures, at each end of the cavity. The patent was granted and published on 5 May 1993 [5].

Whilst heading up a section specialising in payload signal generation and processing for military communications payloads, I was called in to solve a problem that was causing a major headache at the European Space Agency. The Olympus satellite, their flagship project, was exhibiting unstable signals during environmental testing at the spacecraft level. This was eight years into the programme and was presenting a serious schedule delay. Experts from all over Europe had been asked to address the problem, which was identified within the local oscillator equipment, but no solution had been found. I was exiled to the contractor in Belgium, together with a couple of my engineers, and told not to come back until the problem had been solved. We established that the oscillator frequency was not, in fact, determined by the cavity dimensions but by the resonant characteristics of the input circuit. There was a basic design flaw. This was demonstrated by running the oscillator with one end plate removed, much to the amazement of the test technician. This example of the difficulty of matching the impedance of the input circuit to that of the cavity has recurred time and time again for a number of experimenters trying to get an EmDrive cavity working. Microwave resonant cavities can appear to be simple devices, but they require a rigorous design approach, or it will become fiendishly difficult to obtain correct operation. The Olympus lesson remained as a warning against complacency throughout the tuning and testing of subsequent EmDrive thrusters.

A striking example of the benefits that can come from military satellites and the need to protect them occurred when I was asked to manage a rapid update to the NATO 1VB payload for the North Atlantic Treaty Organisation. This was due for launch in 1993. During the Bosnian war, the UHF subsystem required modification to enable forward observers to communicate target information for aircraft strikes. The team literally worked 24 hours a day, 7 days a week, to design, manufacture, qualify and test the necessary equipment. There was a huge relief once the satellite was launched, and the launch party on Cocoa Beach near the Cape Canaveral site in Florida was spectacular. The following day, much hung over, together with our lead engineer, I flew to Sunnyvale, California, to carry out low-earth orbit operations from the famous Blue Cube United States Air Force (USAF) control facility. This included a critical antenna deployment, which was rewarded with a round of applause from observers in the control room when it was successful. Four days after the satellite was declared operational, the first NATO air strikes were carried out, which signalled the beginning of the end of this small, but brutal, war in Europe. Some months later, I requested time on the satellite to carry out performance tests so we could complete the contract and the company could be paid. I was called to a meeting with a Special Air Service (SAS) officer from Hereford, who told me in no uncertain way that he was not going to allow a test to interfere with communications with his guys, who were lying in cold, wet ditches in Bosnia. Apparently, their only

FIGURE 1.6   The author at the NATO 1VB launch.

comfort was their ability to communicate via the satellite, and I was likely to receive some fairly explicit messages if I were to go ahead. We abandoned the tests, and as it all seemed to be working well, the company got paid (Figure 1.6).

While I was pursuing ideas for EmDrive, as a relaxation from the rigours of my day job, Ron Evans, a mathematician and aerodynamicist at BAE systems, was starting up a study of gravitational physics, which grew into the industry-sponsored university project called Greenglow. The study covered all the many propulsion concepts that required new physics. These included gravitomagnetism, Mach effects, and quantum vacuum plasmas. My simple ideas, based on electrical machine theory and microwave physics, must have seemed very dull to those who were reaching out to a warp-drive future. Nevertheless, when Ron eventually wrote his book [6], he mentioned EmDrive, and in the 2016 British Broadcasting Corporation (BBC) horizon documentary about Greenglow, I got to demonstrate some of our equipment. The serious message was somewhat spoiled, however, as in an attempt to warn of the dangers of working with high-powered microwaves and the usefulness of the cut-off condition, the inevitable popcorn cooking in a microwave oven was included.

In 1994, my employer, then Matra Marconi Space, decided to switch my efforts from military satellites to managing high-power TV broadcast payloads on a series of Hotbird satellites for Eutelsat. The budgeting responsibility involved in this task revealed the enormous cost and complexity required to simply move such satellites, which weighed around 3 tonnes at launch, from the transfer orbit achieved by the launch vehicle to the final operational position in a geostationary orbit. I concluded that an EmDrive engine could easily carry out this function, reducing the satellite launch mass by half. While flying to Tokyo to meet with NEC, who were manufacturing some equipment for the Hotbird satellite, I had a chance meeting with Sir Richard Branson in the front of the aircraft. He was on his way to start a trans-Pacific balloon flight, which made being a successful entrepreneur seem very exciting. I suggested to him that he move into the space industry, where there was both excitement and large profits, while he inspired me to think about starting my own company.

However, first of all, I had to offer my invention to my employer. This was part of my employment contract, and if I were to pursue a separate path, I needed to have made the offer. The initial thruster design contained a large-shaped dielectric element, and the geometry of the cavity was carefully designed to provide a good impedance match with the cavity itself. The principle of using the geometry of components to provide an efficient match between microwave devices is well established in the microwave design fraternity. It was therefore with considerable dismay that I learned from a recently appointed director that elementary optical physics did not allow this to happen. The fact that the very communication payload that I was currently responsible for contained 20 critical waveguide loads operating successfully on this same principle did not dissuade him. A lengthy critical report of the whole idea was produced, which was circulated amongst the top echelon of the company, and the offer was soundly rejected.

Ominously, the report concluded with a warning that my career would be in jeopardy if I did not abandon these crazy ideas. This warning was followed up by an attempt to strip me of my Chartered Engineer status by expelling me from the Institute of Electrical Engineers. A friend, who was well respected within the Institute, assured me that this was not going to happen. I think he knew more about the high-level politics within the company than I did because the limited technical competence exhibited by the director soon became apparent in other areas. Within a few months, he had been replaced. A sympathetic French executive then explained that even if the idea did have merit, there was no way that such a disruptive concept would be considered for at least another 10 years. At that time, the Ariane 5 launch vehicle desperately needed a clear run of at least this length of time to cover the investment that had been made. A simple idea that would halve the total launch mass of typical satellites was not what they wanted to hear. As compensation, I was then tasked with the signal and payload design of the proposed Galileo European Navigation Satellite System. It was probably felt that this technical challenge would stop further thoughts of EmDrive. However, in April 1998, I filed a second patent, based on a tapered cavity, and showed how the thruster complied with both conservation of momentum and conservation of energy. This patent was granted on 19 April 2000 [7] and provided the incentive to move on to a new phase of the work, described in Chapter 3. But before the next stage of this tale is told, we must first understand the basic theory of EmDrive.

REFERENCES

(Note. Many of the reference papers can be downloaded from the SPR Ltd website www. emdrive.com)

1. Chevaline: Polaris Improvement Programme. National Archives ref CAB301/734.
2. Cullen A L. Absolute Power Measurement at Microwave Frequencies. IEE monograph No 23. 15 February 1952.
3. Bailey R A. A Resonant-Cavity, Torque-Operated Wattmeter for Microwave Power. IEE monograph No 138R. June 1955.
4. Jennison R C and Drinkwater A J. An Approach to the Understanding of Inertia From the Physics of the Experimental Method. University of Kent. *J. Phys. A: Math. Gen* Vol. 10 No. 2 (1977) 167–179.
5. Shawyer R. Electrical Propulsion Unit for Spacecraft. Patent No GB2229865B.
6. Evans R. *Greenglow and the Search for Gravity Control*. Troubador Publishing 2015. ISBN 978-1784620-233.
7. Shawyer R. *Microwave Thruster for Spacecraft*. Patent No GB2334761B.

2 The Basic Theory

As this is an engineer's tale, any description of EmDrive theory will include equations, diagrams and graphs. These are the tools of our trade. However, the basic physics of EmDrive is quite straightforward, and an understanding of classic mechanics and electromagnetics is all that is required. The actual engineering of an EmDrive cavity is rather more complex, and it appears that actually building a cavity to produce thrust is beyond most experimenters, and thus they abandon the physics, which is a shame.

The heart of the theory is the thrust equation, which is derived from the well-established equation for "radiation pressure":

$$F = \frac{2P}{c} \tag{2.1}$$

where F is the force in Newtons, P is the power of the electromagnetic beam in Watts, and c is the speed of light in metres per second.

The equation gives the **force** of an electromagnetic beam being totally reflected from a surface in free space. It is in fact an equation for force, not **pressure**, which is probably one of the first reasons for initial confusion.

However, we are looking for the force created by the reflection of an electromagnetic wave from the end plate of a cavity. A microwave cavity can be thought of as a waveguide with end plates. The velocity of the electromagnetic wave along the waveguide is called the guide velocity and is designated V_g. It is determined solely by the properties of the waveguide and the mode of transmission. However, the effect on the force created by the reflection of the end plate is the most important characteristic. In free space, the velocity of propagation is c, and equation (2.1) holds true. In a waveguide, the velocity is always lower than c, and the reflection force is lower. The reduction in force is in direct proportion to the ratio of guide velocity to c.

Thus, the force at end plate 1 can be given by:

$$F_1 = \frac{2P}{c}\left\{\frac{V_{g1}}{c}\right\} \tag{2.2}$$

The concept of reflection force being proportional to propagation velocity is perhaps more easily understood by reverting to the physicist's picture of particles rather than waves, with photons bouncing off the end plates and therefore the pressure increasing as the velocity increases. This simple picture brings to mind ball bearings fired from a gun and, unfortunately, can also be deeply misleading. It is much safer to stick with wave theory and equations.

Equation (2.2) infers that if the guide velocity is different at each end of the cavity, then a force difference will be set up, which can be given by:

DOI: 10.1201/9781003456759-2

$$T = F_1 - F_2 = \frac{2P}{c}\left\{\frac{V_{g1}}{c} - \frac{V_{g2}}{c}\right\} \qquad (2.3)$$

where T is defined as the cavity thrust in Newtons.

One property of a waveguide that determines the guide velocity is its width. The smaller the width, the slower the velocity. Taken to one extreme, i.e., an infinite width, the velocity is c, as this is equivalent to free space. Of more interest is what happens as the width gets smaller. For a given frequency of operation, the width of the waveguide reaches a minimum, referred to as the cut-off condition. At this point, the velocity goes to zero. It is this effect that allows us to use microwave ovens and see the contents without getting cooked ourselves. The holes in the metallic mesh, forming part of the door, are so small that they are operating well into the cut-off condition so that no radiation can pass through.

With the dependence of V_g on the waveguide width established, it becomes easy to visualise a cavity where the velocities at each end plate are different. The cavity simply needs to be a truncated cone or pyramid.

It is important to understand that the forces due to the reflection of an electromagnetic wave are actually a result of the same electromagnetic phenomena that occur in a standard electric motor. In his original paper, Cullen derives equation (2.1) from the basic electrical machine equation.

$$F = BLi \qquad (2.4)$$

where F is the force in Newtons, B is the flux density in Tesla, L is the length of the conductor in metres and i is the induced current in amps.

This means that the force is being produced **by** the end plate, i.e., the conductor, and is not a force being applied **to** the end plate. This is where the picture of photons merely hitting the end plate causes confusion. The electromagnetic wave interacts with the conductor to produce a force in exactly the same way as in an electric motor. The classic criticism of EmDrive operation being like pushing on the windshield of a car from the inside, as shown in Figure 2.1, is not correct. Clearly, any such force F would be opposed by a reaction force R.

Professor Yang Juan, who is a world expert in microwave ion engines, described EmDrive as *"a black box pushing against the rest of world, just as we are"*. She was having her own problems with criticism from the serried ranks of Chinese academics, and we will meet her later in the story.

We now have a box that is producing a force with nothing else coming out of it. How can this comply with the law of conservation of momentum?

At this point, it is important to realise that an electromagnetic wave carries momentum. Although the wave has no mass, it has velocity and energy. Einstein has told us that mass and energy are equivalent; therefore, energy and velocity will give momentum. In the cavity, the electromagnetic wave bounces backwards and forwards between the end plates, gradually giving up its momentum and thus producing forces at each end plate. Just as the force at the larger end plate F_1 is greater than the force at the smaller end plate F_2, the momentum given up at the large end plate is greater than the momentum given up at the small end plate.

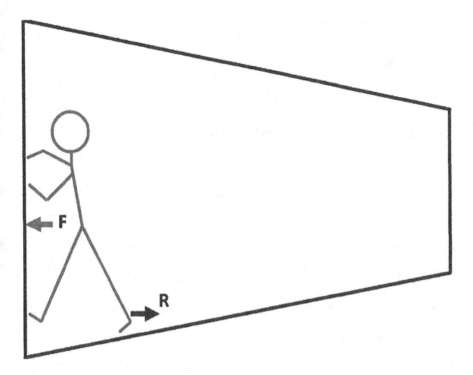

FIGURE 2.1 A wrong idea.

However, in any free-moving system, momentum must be conserved. The net momentum given up by the electromagnetic wave must be balanced by the momentum gained by the cavity as it accelerates. Thus, the direction of the acceleration is opposite to the direction of the net momentum of the electromagnetic wave and therefore the direction of thrust T. This is a direct result of Newton's third law, where an action, the cavity pushing in one direction, gives a reaction, the cavity accelerating in the opposite direction. If you don't believe this, try pushing against a brick wall; your body will accelerate in the opposite direction, and you will fall over. Note that in a typical description of a rocket motor, the direction of acceleration is usually shown as the same as the direction of thrust. Actually, it should be described as the same direction of the reaction force due to the thrust of the rocket exhaust mass. EmDrive is a reaction machine in just the same way as a rocket. Newton's laws do not allow the existence of a reactionless machine.

Conservation of momentum can be confirmed by carrying out a simple balance of electromagnetic and mechanical momenta for the classic Photon Rocket concept and comparing it with a similar balance for an EmDrive thruster. Figure 2.2 shows a simple diagram of a photon rocket, sometimes referred to as a flashlight rocket, as it resembles the bulb and reflector of a flashlight.

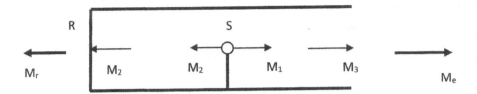

FIGURE 2.2 Momentum diagram for a photon rocket.

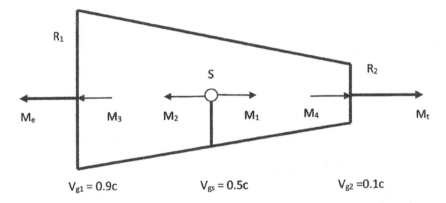

FIGURE 2.3 Momentum diagram for an EmDrive thruster.

Assume source S produces electromagnetic beams with momentum M_1 and M_2, and reflecting plate R reflects the beam with momentum M_2 transferred to the reflector. Assume S and R are mechanically connected and there are no losses.

Assume the momentum vector convention is $\rightarrow$ is positive.

Then total electromagnetic momentum $M_e = M_1 - M_2 + M_3 = M_1$

Momentum transferred to the rocket $= M_r$

By conservation of momentum $M_r + M_e = 0$

Then $M_r = -M_1$.

Clearly, the rocket's momentum is in the opposite direction of the momentum of the emitted electromagnetic beam.

If we now consider the momentum balance for an EmDrive thruster, as illustrated in Figure 2.3, things are a little more complex.

Assume source S produces electromagnetic beams with momentum M_1 and M_2, and reflecting plates R_1 and R_2 reflect the beams with momentum M_3 and M_4.

Note that the guide velocity decreases as the beam moves down the axis of the cavity in accordance with the decreasing width, and therefore the momentum will also decrease.

Assume that in free space, the absolute momentum of the beam is M_0. Assume the guide velocities V_{g1} at R_1 and V_{g2} at R_2 produce absolute momenta of $M_3 = 0.9\,M_0$ and $M_4 = 0.1\,M_0$, and that the guide velocity V_{gs} at S produces absolute momenta of $M_1 = M_2 = 0.5\,M_0$.

Then, for one complete transit of the cavity, applying the vector convention, the total electromagnetic momentum

$M_e = M_1 - M_2 - M_3 + M_4$

Therefore, $M_e = 0.5\ M_0 - 0.5\ M_0 - 0.9\ M_0 + 0.1\ M_0 = -0.8\ M_0$

Momentum transferred to the thruster $= M_t$.

By conservation of momentum, $M_t + M_e = 0$

Therefore, $M_t = 0.8\ M_0$.

Note that the thruster momentum is in the positive direction, i.e., large end plate to the small end plate, because the total electromagnetic momentum is in the negative direction. This supports the idea of the cavity pushing in the negative direction, resulting in the cavity accelerating in the positive direction. The statement that EmDrive contravenes the conservation of momentum is clearly false.

But the cavity has nothing to push against, you may say, as did many people at the beginning of the twentieth century when rockets operating in the vacuum of space were proposed. In 1920, the New York Times claimed that a rocket would not work in space. A rocket has nothing to push against, they wrote. It doesn't matter. As a 1969 retraction article finally acknowledged. The rocket relies on Newton's laws to comply with the Law of the Conservation of Momentum. So it took almost 50 years for the New York Times to understand the conservation of momentum. It's not surprising that many people still don't understand EmDrive.

As was pointed out in Chapter 1, discussions as to whether EmDrive is a closed or open system have been the basis of much of the controversy surrounding the technology. Whereas it is straightforward to determine the open or closed nature of purely mechanical or electrical systems, once electromagnetic waves are introduced, the question becomes more complicated. This is due to the need to consider Einstein's theory of special relativity. The starting point in special relativity is that the velocity of an electromagnetic wave in free space is always c, no matter what the velocity of the device producing the wave is. It is generally said that the speed of light is constant, which is a fact that has been proven over and over again since it was first proposed. The speed of light is also considered to be the maximum possible speed for any wave or particle, and special relativity and the subsequent general theory of relativity have been the cornerstones of modern physics. Indeed, equation (2.1) can be derived from Einstein's famous $E = Mc^2$, which can even be found inscribed on T-shirts, so it has to be accepted. However, when an electromagnetic wave is constrained within a waveguide with end plates, thus forming a closed cavity, is the cavity and its source of microwave power a closed system?

We have seen that the guide velocity is dependent only on the waveguide dimensions and the value of c. However, special relativity states that c is independent of the velocity of the source, so the guide velocity must also be independent of the velocity of the source. Thus, for example, if the cavity were to be accelerated up to half the speed of light, the guide velocity would not be doubled or halved, as c would remain stubbornly constant. Thus, the performance of the cavity is independent of the velocity of the cavity itself. Note that during the acceleration, a Doppler shift will occur between the frequency of the incident and reflected waves at one end plate. This shift will not be fully corrected when it reaches the other end plate, as by then the cavity velocity will have increased. The problem of Doppler shift in high-Q cavities will be addressed later when superconducting cavities are discussed. However, it seems that the forces and physical laws that occur outside the cavity have an effect inside the cavity, and thus EmDrive could reasonably be considered an open system.

Inspection of equation (2.1) gives rise to another difficulty with radiation pressure; it is incredibly small. The value of c is 300,000 km/s (3×10^8 m/s), so even with a power of 1 kW, the force produced is a mere 7 micro-Newtons (7×10^{-6} N). However, the saving grace of an EmDrive thruster is the so-called Q factor of the cavity. This can be simply thought of as the number of times a full-power electromagnetic wave moves backwards and forwards along the cavity axis. The Q factor in any resonant system has a multiplicative effect on the forces present. Thus, if the frequency of the marching feet of an army matches the natural resonant frequency of a bridge and if the bridge has a significant Q value, then the forces generated will tear the bridge apart. Fortunately, the engineers who design bridges are very familiar with this concept and will include structural damping, which will ensure the Q value is well below the required safety factor.

However, in a well-designed microwave cavity, manufactured to very precise dimensions and properly fed from a microwave source of the correct frequency, Q values of 50,000 (5×10^4) can be readily achieved. This has uncooled, silver-plated internal surfaces. Thus, the force generated at an end plate can be 50,000 times the 7 micro-Newtons for one reflection, and the resulting 350 milli-Newtons (3.5×10^{-1} N) for one kilowatt of power is a useful thrust for space applications. In high-energy physics, linear and circular particle accelerators are common. In these accelerators, superconducting microwave cavities with Q values up to 5×10^9 are lined up in their thousands to produce very high particle velocity. Clearly, if this technology, which is readily accessible in today's world, can be applied to an EmDrive cavity, then the 350 milli-Newton force can be multiplied up to 35 kN (3.5×10^4 N). This is 3.5 tonnes of force, a very useful thrust in tomorrow's world.

From equation (2.3), the basic equation for thrust for an EmDrive cavity will therefore become:

$$T = \frac{2PQ}{c} \left\{ \frac{V_{g1}}{c} - \frac{V_{g2}}{c} \right\} \tag{2.5}$$

There are, of course, many engineering problems on the way, and the rest of this book will guide the reader through the theoretical and experimental solutions to the problems. However, it is first necessary to address the basic engineering theory of a simple microwave cavity.

We have seen that the cavity must resonate to build up sufficient Q so that the forces will be useful. A fundamental property of the electromagnetic wave, its wavelength, is now introduced. An electromagnetic wave in free space comprises a magnetic field H and an electric field E, at 90° to each other and varying in amplitude in a sinusoidal manner. This field structure is moving in a direction 90° to the field directions at velocity c. If a movie of the field structure were made, it would produce a simple sinusoidal wave. This wave repeats itself when it has travelled a distance of one wavelength, designated λ_0 in free space.

Thus

$$\lambda_0 = \frac{c}{f} \tag{2.6}$$

where f is the frequency of the electromagnetic wave in Herz. However, in a waveguide, things are rather more complex. The wavelength of the energy transmitted down the axis of the waveguide is the guide wavelength, designated λ_g. We have already come across the concept of cut-off when the dimensions of the waveguide become too small for an electromagnetic wave to be transmitted. The wavelength at cut-off is designated λ_c, and the value of λ_c is determined by the width of a rectangular waveguide designated a, where for a TE_{10} mode:

$$\lambda_c = 2a \tag{2.7}$$

The value of λ_c in both rectangular and circular waveguides is dependent on the dimensions and the mode of transmission, which we will consider later. The relationship between λ_g, λ_0 and λ_c in a vacuum-filled waveguide is given by:

$$\lambda_g = \frac{\lambda_0}{\sqrt{1 - \left(\frac{\lambda_0}{\lambda_c}\right)^2}} \tag{2.8}$$

To ensure that a microwave cavity resonates and provides the necessary Q, the electrical length of the cavity must equal an integer number of half-wavelengths. This will ensure that the E and H fields travelling in opposite directions, as they move up and down the waveguide, add at each point. This is easily accomplished if the guide wavelength is constant along the axis of a waveguide of constant width or diameter. However, we have seen that an EmDrive cavity is a truncated pyramid or cone, where the width or diameter is continually changing. Thus, to design a resonant EmDrive cavity, a method of numerical analysis must be adopted. This is where the stalwart of engineering design, a spreadsheet, can be readily employed. For a given frequency, maximum and minimum widths or diameters, and mode of transmission, the guide wavelength can be determined from equation (2.8) for increments along the axial length of the cavity. The incremental wavelengths are integrated to produce the required number of half-wavelengths at the resulting physical length of the cavity. This is illustrated in Figure 2.4, where a resonant cavity of three half-wavelengths is shown.

As can be determined from equation (2.8), the incremental guide wavelength does not change linearly with waveguide dimension. As the dimension narrows and approaches cutoff, the guide wavelength rapidly approaches infinity as the velocity approaches zero. This is plotted in Figure 2.5, where the axial length at the small end plate end is 161.5 mm. The effect can also be seen in Figure 2.4 by the stretching of the third half-wavelength as the small end plate is approached.

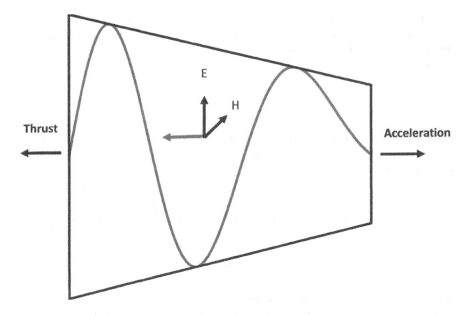

FIGURE 2.4   EmDrive cavity of length three half-wavelengths.

Figure 2.5 also illustrates the criticality of the dimensions of the cavity, as it is important to ensure cut-off is not reached; otherwise, the travelling wave stops and EmDrive ceases to function. The temptation and ease of using a commercial finite element software package to design the cavity should be balanced by the fact that some versions completely ignore cut-off conditions and can give misleading designs. A very senior UK government scientist dismissed them as *"Mickey Mouse programmes, giving pretty pictures but not correct in real life"*. This seemed a bit harsh, but I have found that a high-resolution spreadsheet analysis, typically using 0.1 mm increments and using equations from the real world, has consistently given accurate designs when compared with test results.

We have also seen from equation (2.8) that the value of the guide wavelength λ_g is dependent on the cut-off wavelength λ_c, and we have stated that λ_c is dependent on the dimensions and the mode of transmission. We must therefore now take a brief look at transmission modes. For waveguides and cavities, they are divided into transverse electric (TE) and transverse magnetic (TM) modes. By convention, in a TE mode, the electric field is always transverse to the direction of propagation, along the z axis. Thus, in a rectangular waveguide, it is directed across the width, whose dimension is designated a. The height of a rectangular waveguide is designated b. For TM modes, the electric and magnetic fields are rotated 90° about the z-axis. Mode designations include two subscripts, m and n. For rectangular waveguides, the designations become TE_{mn} or TM_{mn}, whereas for circular guides, they are TE_{nm} or TM_{nm}. These subscripts indicate the number of half-sine wave variations of field components in the x and y directions, respectively.

The number of possible modes is infinite, as m and n can be any integer from zero to infinity. However, the higher-order modes are limited by the cut-off

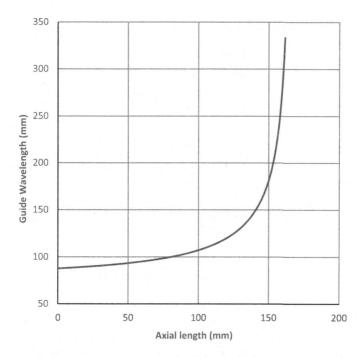

FIGURE 2.5   Guide wavelength plotted against axial length.

wavelength. For any mode in a rectangular waveguide, the cut-off wavelength can be determined by:

$$\lambda_c = \frac{1}{\sqrt{\left(\dfrac{m}{2a}\right)^2 + \left(\dfrac{n}{2b}\right)^2}} \tag{2.9}$$

For circular waveguides, the determination of the cut-off dimension is quite different. The diameter at which cut-off occurs can be determined by:

$$d_c = \frac{\lambda_0}{K_{nm}} \tag{2.10}$$

Table 2.1 gives the values of K_{nm} for lower-order TE modes in a circular waveguide. The derivation of K_{nm}, together with a full description of waveguide modes, is given in [1].

When a waveguide is transformed into a resonant cavity by fitting two end plates at a distance equal to an integer number of half wavelengths, that number is designated p and appears as a third subscript. Thus, a TE_{113} mode operating in a circular cavity will have a cavity wavelength of 1.5, i.e., 3 half wavelengths. From Table 2.1, TE11 is seen to have the highest value of K_{nm}, and therefore this mode will yield the smallest cut-off diameter. However, all engineering designs are compromises, and the smallest cavity diameter and shortest length are not always the best solutions.

TABLE 2.1

Values of K_{nm}.

nm	1	2
0	0.82	0.448
1	1.706	0.589
2	1.029	0.468

The Q of the cavity is essentially determined by the volume of the cavity, though the theoretical value for a circular cavity is given by:

$$Q_t = \frac{\left(\dfrac{377}{2R_s}\right)\left[14.684 + \left(\dfrac{1.571pd_m}{L_0}\right)^2\right]^{1.5}}{14.684 + \left[\dfrac{d_m\left(\dfrac{1.571pd_m}{L_0}\right)^2}{L_0}\right]} \tag{2.11}$$

where R_s is the surface resistance of the cavity material at the resonant frequency, L_0 is the length of the cavity and d_m is the mean diameter.

Before we leave the consideration of cavity modes and dimensions, a very important point must be made about the detailed geometry. The theory assumes that the geometry and machining tolerances during cavity manufacture are such that at any point across the wavefront, the path length between reflection points on each end plate is equal. We have said that a simplified view is that the Q of the cavity equals the number of times that a full-power wavefront traverses the cavity axis. For a Q value of 50,000 and a path length of 160 mm, this implies an accuracy requirement of better than three microns is required to maintain equal path lengths across the wavefront. In practice, the relationship between machining tolerances and Q is very complex, and with correct plate alignment, a cavity can achieve this Q value with a specified machining tolerance of 0.1 mm.

To achieve such accuracy, the end plates clearly cannot be completely flat, as the path length difference between the side walls and the central axis would be excessive and lead to an unacceptably low Q value. A number of circular cavity geometries with equal path lengths across the wavefront have been designed and tested, each giving satisfactory Q levels. The simplest concept is shown in Figure 2.6. This shows a convex small plate and a concave large plate, with all path lengths being equal fractions of the same radius R_1.

A geometry that is easier to manufacture was chosen for the flight thruster and is shown in Figure 2.7. This shows the central part of this cavity is cylindrical, with a diameter equal to the diameter of the small end plate. The tapered annular section terminates in a curved large end plate section with a radius R_2. The central section of

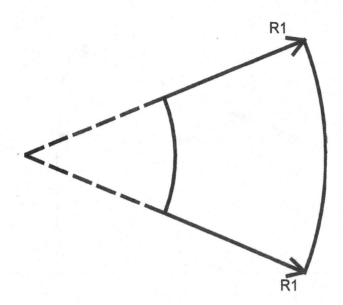

FIGURE 2.6 Simple cavity geometry for equal path lengths across the wavefront.

the large end plate and the full area of the small end plate are therefore both flat. With this geometry, it was also found to be easier to align the two end plates to maximise Q.

A third geometry, shown in Figure 2.8, was used for a superconducting cavity, which enabled the large end plate to be flat. However, this meant that the concave small end plate needed a complex shape, which was designed and machined using high-accuracy, numerical techniques to ensure equal path lengths were maintained across the wavefront.

A further major theoretical aspect of EmDrive is how to actually persuade microwave power to propagate inside the cavity. It is quite easy to set up a small signal wave propagation system with any type of input antenna, such as a slot, a probe or a loop. However, to transfer power from the antenna to the very high-Q cavity, there must be a very close impedance match between the cavity and the antenna. The first problem is that the cavity impedance depends on the resonant frequency, dimensions and mode, and it changes along the axis of the cavity. For a circular cavity, the cavity wave impedance can be calculated from:

$$Z_w = 376.7\sqrt{1 - \left(\frac{\lambda_0}{dK_{nm}}\right)^2} \qquad (2.12)$$

where d is the diameter of the cavity at the axial position. The wave impedance is plotted in Figure 2.9, where the rapid decrease in value can be seen as the axial length increases and the small end plate is approached. At the cutoff diameter, the impedance falls to zero.

The input circuit must therefore be designed to have the correct impedance match at the position chosen along the cavity length and must maintain that match over the

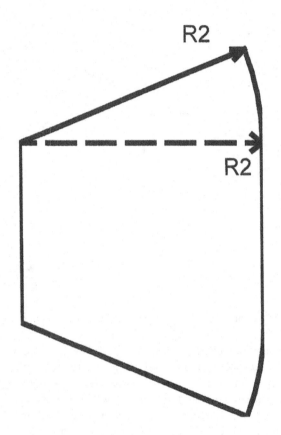

FIGURE 2.7   Flight thruster cavity geometry.

significant temperature rise of the antenna during power transfer. The temperature rise is due to the high circulating currents in the antenna itself, which must be resonant at the input frequency and have a Q value that matches that of the cavity. The input circuit must therefore comprise both the antenna and a very sensitive tuning mechanism. Also, the actual input frequency must inevitably be continuously tuned, using a closed-loop frequency control system, to match the changing resonant frequency, as the cavity temperature changes with input power level and the environmental temperature. As has been mentioned in Chapter 1, the Olympus satellite programme showed that it is very easy to assume your cavity is resonating with the input power when, in practice, it is simply the input circuit that is absorbing all the power.

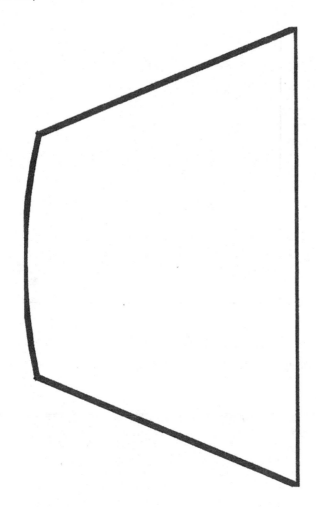

FIGURE 2.8 Superconducting cavity geometry.

This chapter has hopefully shown that although the physics of EmDrive is simple, building a working EmDrive cavity requires considerable attention to detail. A number of attempts that have failed have led to claims that EmDrive can't possibly work. These claims were described by one science journalist, David Hambling, as being as if, in the early days of flight, someone had tried to copy the Wright Flyer but had built the copy with the wings upside down. They then found that it wouldn't fly and concluded that aeroplanes were impossible. I would merely add that if you wish to build a working aeroplane, first learn the theory of flight. If you wish to build a working EmDrive, first learn the theory of EmDrive.

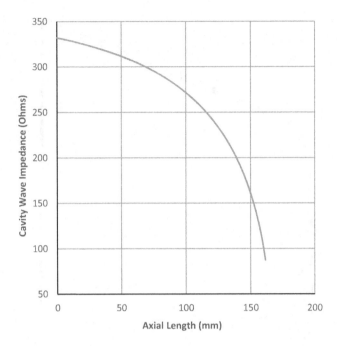

FIGURE 2.9   Cavity wave impedance plotted against axial length.

REFERENCE

1. Rizzi P A. *Microwave Engineering Passive Circuits.* Prentice Hall 1988. ISBN 0-13-586702-9 025.

3 The Experimental Engine

In May 2000, I decided to investigate setting up my own company to carry out further research and development on EmDrive. In this, I was greatly helped by Charles Dawes of Inventions Direct Ltd., who had a background in supporting new inventors. A detailed review of the early documentation was carried out by both technical and patent experts, and it was felt that the company should include a number of shareholders who could contribute technical, commercial and financial expertise. I was very grateful to leave the financial running of the company to Mike Sheridan, a Chartered Accountant with considerable experience in both large and small companies, and on October 20, 2000, Satellite Propulsion Research Ltd. (SPR Ltd.) was set up with myself as director and Mike as company secretary. Mike's European contacts were immediately helpful, and a day trip to Paris to meet up with a French banker was the start of many funding prospects that were followed up on in the early days of the company. The original four shareholders were soon joined by Dr. Richard Paris of Abertay University in Dundee. Richard was a reader in mathematics and a specialist in Magneto Hydrodynamic (MHD) theory, with considerable experience in space propulsion and nuclear physics. Richard carried out a detailed review of the mathematics behind the theory of EmDrive, and this has ensured that the math has withstood the rigorous scrutiny and review that it has received over many years. Over the next decade, SPR Ltd. acquired additional shareholders to add marketing and funding capabilities to the company.

With SPR Ltd. beginning to make progress, I left what had now become Astrium in mid-February 2001. I had funds for the next two years from a Marconi shareholding, which I had been given as part of my salary. However, I was about to face the first major problem for my fledgling business. When I left, the share price was around £12.50; by September 2001, it had fallen to 29p. Whilst this was a major blow to the UK telecoms business, it was a disaster for me, and a new source of funds was urgently sought. Our first major problem was soon followed by an amazing piece of luck. The UK government had recently announced its Smart awards scheme, and an application from SPR Ltd. was quickly on its way to the Government Office for the South East, which was managing the scheme locally. By the 3rd of July, we had gratefully accepted a grant for our proposed feasibility study, which was officially started on the 1st of August. We now had funding for the next 18 months.

A couple of months later, the Department of Trade and Industry (DTI) issued the following press release:

SATELLITE TECHNOLOGY WINS AWARD FOR THE EMSWORTH TEAM
Work on satellite propulsion technology has led to a West Sussex firm being singled out for a prestigious innovation award.
Satellite Propulsion Research Ltd, located in Emsworth, has received the £45,000 award from the DTI's Small Business Service (SBS) to continue feasibility studies into the project. It will develop and test an experimental engine using microwave

DOI: 10.1201/9781003456759-3

technology to convert solar energy directly into engine thrust in the later stages of the launch. It would lead to significant reductions in the weight of the satellite, and its lifetime could be extended considerably.

The technology could also be used as the main propulsion for deep space missions, where small levels of thrust over many months would enable low cost spacecraft to deliver complex payloads. The first experimental engine has been built and initial tests confirm the operating principles.

As the press release implied, the experimental engine had been built and tested before the award proposal was submitted. The theory and test data were reviewed by both government and industry reviewers, including the UK Ministry of Defence (MoD) at Farnborough and Plessey at Roke Manor Research. A typical comment was that no flaw could be found in the theory, but the test results could be due to any number of spurious forces, and the engine should be subject to a variety of tests aimed specifically to calibrate or remove these forces.

The conceptual design of the experimental thruster is given in Figure 3.1. It uses a tapered circular waveguide and a cylindrical dielectric-filled section, resulting in dissimilar wave velocities at each end of the waveguide. Thus, the rate of change of momentum as the wave is reflected from each end wall is different, and hence different forces, designated F1 and F2, are produced. The waveguide assembly is designed to be resonant at the operating frequency, and therefore the force difference is multiplied by the Q of the assembly, producing a useful resultant force on the thruster.

Note that λ_d, the wavelength for an unbounded dielectric, in the dielectric section is smaller than λ_0, the wavelength in free space for the tapered section. But more importantly, the velocity is also lower in the dielectric section than in the tapered section.

This is because of equation (2.6) in Chapter 2:

$$\lambda_0 = \frac{c}{f}$$

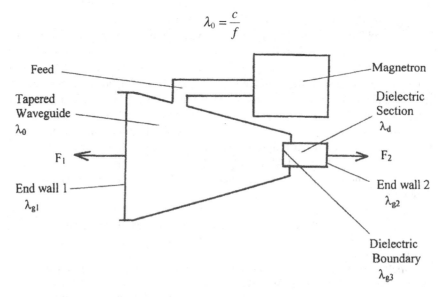

FIGURE 3.1 Conceptual design of the experimental engine.

and from basic electromagnetic theory, for any insulating material:

$$\lambda_d = \frac{\lambda_0}{\sqrt{\mu R \varepsilon R}} \tag{3.1}$$

where μR is the relative permeability and εR is the relative permittivity of the material.

For low-loss microwave dielectric material, μR is 1.

Therefore, from equations (2.6) and (3.1),

$$\lambda_d \frac{c}{f\sqrt{\varepsilon R}} \tag{3.2}$$

The velocity at the end wall of the dielectric section is therefore:

$$V_{g2} \frac{c\lambda d}{\lambda_{g2}\sqrt{\varepsilon R}} \tag{3.3}$$

where λ_{g2} is the guide wavelength throughout the dielectric. This is constant because the diameter of the dielectric section is constant.

The equation for thrust when a dielectric section is included in the cavity is therefore:

$$T = \frac{2PQu}{c}\left\{ \frac{\lambda_0}{\lambda_{g1}} - \frac{\lambda d}{\lambda_{g2}\sqrt{\varepsilon R}} \right\} \tag{3.4}$$

The guide wavelength in the tapered section at the air/dielectric boundary is λ_{g3}. It is important that at this boundary, the wave impedance in the air section matches that within the dielectric section. The experimental cavity was designed to operate in TM013 mode, and the wave impedance in TM mode for the air/dielectric boundary can be calculated from the following equations:

The impedance of air

$$Z_3 = \frac{\lambda_0}{\lambda_{g3}}\sqrt{\frac{\mu_0}{\varepsilon_0}} \tag{3.5}$$

The impedance in dielectric

$$Z_2 = \frac{\lambda d}{\lambda_{g2}}\sqrt{\frac{\mu_0 \mu R}{\varepsilon_0 \varepsilon R}} \tag{3.6}$$

where μ_0 is the permeability of free space and ε_0 is the permittivity of free space.

The cavity must clearly be designed such that $Z_3 = Z_2$.

In practice, the cavity was designed so that the position of the dielectric section along the axis of the cavity could be varied to allow different modes to be propagated and so that fine-tuning of the impedance match could be achieved.

The dielectric properties also had a major effect on the unloaded Q of the cavity Q_u. This is the Q value attributed to the cavity itself, unloaded by the input circuit, with the stored energy not being reduced by kinetic energy transfer due to acceleration. Importantly, it is the Q value used in the equation for thrust (3.4). In the experimental cavity:

$$\frac{1}{Q_u} = \frac{1}{Q_a} + \frac{1}{Q_d} \tag{3.7}$$

where Q_a is the unloaded Q of the air section and Q_d is the unloaded Q of the dielectric section. Note that the cavity Q value measured during testing is the value measured with the cavity loaded by the input circuit. Thus, Q_L can be calculated from:

$$\frac{1}{Q_L} = \frac{1}{Q_u} + \frac{1}{Q_i} \tag{3.8}$$

where Q_i is the Q value of the input circuit. For a correctly designed input circuit, carefully tuned to match the input circuit impedance to the wave impedance at the input position:

$$Q_u = Q_i. \text{ Thus } Q_L = \frac{Q_u}{2} \tag{3.9}$$

Figure 3.2 gives a diagram of the actual experimental engine. The engine uses a magnetron as a source of microwave energy, which is fed to the thruster through a tuned waveguide assembly. The electromagnetic wave is launched from a slot in the side wall of the thruster and propagates with an increasing guide wavelength towards the dielectric section. As the circular waveguide cross section decreases, the waveguide impedance also decreases, such that the impedance at the dielectric boundary is the same on both sides. This ensures propagation continues within the dielectric section without reflection at the boundary. The thruster was designed around a Siemens dielectric resonator type LN89/52B with a dielectric constant εR of 38. An operating frequency of 2450 MHz was selected to allow a commercial 850 W oven magnetron to be used as a power source. A rectangular feed structure was used, with a resonant slot selected for the input launcher and a small spacing between the feed waveguide and slot to provide a choke. This allowed reflected power to leak into the EMC enclosure when operating with an unmatched input impedance. Input matching was achieved through feed length adjustment and a tuning screw in the input waveguide. Reflected power was measured by a load mounted on the inside wall of the enclosure. Thruster resonance tuning was carried out by varying the position of the end wall using screw adjusters. Optimal impedance matching at the dielectric boundary was achieved by fine adjustment of the axial position of the dielectric section. The magnetron is cooled by a DC fan, and the enclosure contains sufficient cooling air volume to enable the magnetron to be run for 50-sec periods without overheating. Two thermistors are used for temperature monitoring, and a thermal cutout is used to provide a high-temperature warning.

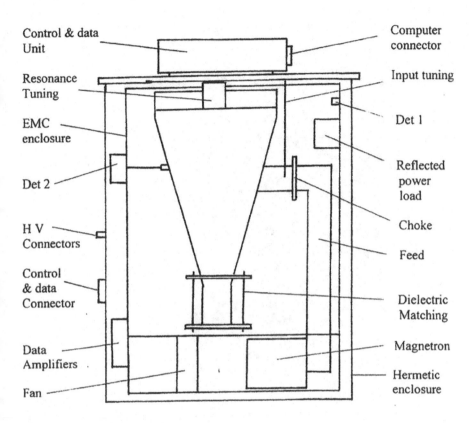

FIGURE 3.2 Diagram of an experimental engine.

A view of the experimental engine without the EMC shielding is shown in Figure 3.3. The primary function of the EMC enclosure was to provide safety from high-power microwave leakage, which, if not shielded from the operator, can cause serious health hazards. Although the EMC enclosure provides initial screening as well as containment of the cooling air, it was necessary to house the data monitoring electronics in separate screened boxes. Also, to reduce thermal buoyancy effects during test runs, a second 'hermetic' enclosure was manufactured. (Note that although this significantly reduced the leakage of heated air, it could not be considered truly hermetic.) Two power detectors were developed, one mounted inside the module enclosure (Det 1) and one mounted inside the thruster itself (Det 2). Det 1 measures the power reflected back into the enclosure due to any mismatch at the thruster/feed interface. Det 2 measures the power level developed inside the thruster's resonant cavity. Both detectors are of the voltage probe type with attenuators and germanium diodes. The probe length and attenuator values were optimised to keep the detected signals within the predicted dynamic range while minimising the risk of arc damage.

Note that very high field strengths exist within the cavity, and therefore high levels can leak from the choke. It was necessary to mount Det 2 in a secondary cavity to prevent arcing. The detector signals are amplified in two stages with offset and gain control.

FIGURE 3.3 Experimental engine without EMC shielding.

Passive integrator circuits are used to provide DC outputs and minimise noise. Note that the raw detector signals consist of 50 Hz pulses of power, as the magnetron itself is powered from a half-wave rectified high-voltage power supply running at mains frequency. The battery to supply the cooling fan, along with amplifier supplies and controls, are mounted in a separate control and data unit, as shown. The data signals from the power detectors and thermistor, together with thrust data from the load cell or precision balances, are converted to digital format and logged on a laptop computer, where they can be stored and processed in Excel spreadsheets. The overall geometry was defined by building a mathematical model of the thruster based on an Excel spreadsheet. The physical length was divided into 0.5 mm sections, and the guide wavelength was calculated for each section. The electrical length for that section was calculated, and the sum of the section electrical lengths was calculated. Thus, variations of diameters, lengths and ε_R could be modelled, with the target of achieving an overall electrical length of an integer number of half wavelengths at the nominal operating frequency. The model also allowed the operation to be modelled in TE11 and TE12 modes, the nearest unwanted modes. The design was optimised to avoid the possibility of any unwanted mode of operation. Once the main parameters had been determined and a suitable commercially available dielectric resonator identified, a tolerance analysis was carried out. The

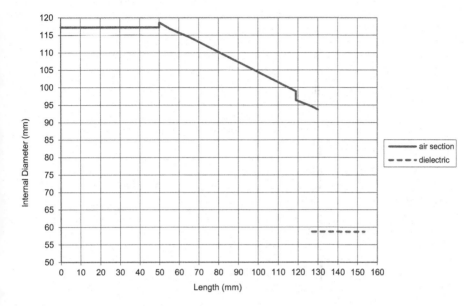

FIGURE 3.4   "As built" dimensions of the experimental cavity.

TABLE 3.1
Predicted and measured resonance lengths

Frequency (MHz)	Resonance Length (mm)	
	Predicted	Measured
2,451.113	37.5	37.8
2,432.113	32	31.9

expected variations in operating frequency, mechanical dimensions and electrical properties were input into the model. This enabled the adjustment ranges for resonance tuning and dielectric matching to be determined. After manufacture, the "as built" dimensions were input into the design model, and this became the reference model against which the operation of the thruster could be compared. The "as built" data is given in Figure 3.4.

The results of the mathematical model were compared to the parameters measured during a set of small signal tests using two narrow-band 100 mW crystal sources operating at 2,451.113 MHz and 2,432.113 MHz. The predicted and measured resonance lengths at these two frequencies are given in Table 3.1.

As can be seen, the measured lengths are well within the 0.5 mm resolution of the model. Future versions of the model increased the resolution to 0.1 mm with increased accuracy of prediction. During these resonance tests, the output from the internal cavity power detector Det2 was plotted both with and without a copper disc short-circuiting the air/dielectric boundary. The results are shown in Figure 3.5.

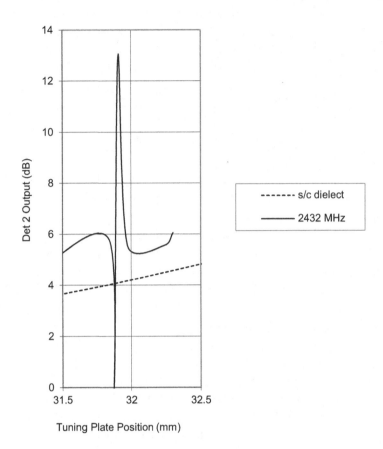

FIGURE 3.5 Tuning plot for 2,432.113 MHz source.

With the short-circuit disc in place, no resonance was measured, clearly showing that the resonance was due to propagation through both the air and dielectric sections of the thruster. An estimate of the Q of the thruster was obtained by measuring the 3 dB bandwidth of the resonance peak in Figure 3.5. At this early stage, our test equipment was pretty basic. The SMART award did not cover the purchase of the normal equipment seen in a microwave laboratory. Swept signal sources, frequency counters, high-power amplifiers, and network analysers easily run up bills approaching twice the value of the award. This meant that both high-power sources and small signal sources had to be fixed-frequency devices, and the normal Q measurement technique had to be modified. Rather than sweeping the frequency to obtain a resonance plot, the cavity length was swept, resulting in Figure 3.5. The equivalent resonance frequency was calculated from the length, and the 3 dB bandwidth was calculated. This bandwidth, which is the frequency difference between the half power points of the central peak of the response, can then be used to calculate the Q value from the equation:

$$Q_L = \frac{F_0}{F_2 - F_1} = \frac{\text{Resonant Frequency}}{\text{3dB Bandwidth}} \tag{3.10}$$

This gave an approximate Q_L value of 5,900. The specified Q of the dielectric section (Q_d) was 14,987, and the calculated Q of the air section from equation (2.11) (Q_a) was 60,562. Thus, from equations (3.7) and (3.8), the predicted Q_L was 6,008. Once again, the measured parameters of the experimental cavity were in agreement with the design predictions. However, the acid test would come when thrust measurements were made.

The expected thrust at the nominal operating frequency of 2,450 MHz could be predicted from equation (3.4), where the following values were assumed:

$$P = 850 \text{ W} \qquad\qquad \mathcal{E}_R = 38$$
$$\lambda_{g1} = 0.20182 \qquad\qquad \lambda_{g2} = 0.02055$$
$$\lambda_0 = 0.12236 \qquad\qquad \lambda_d = 0.01985$$

Then predicted thrust $T = 15.3$ mN (1.56 gm). Note that equation (3.4) is the basic thrust equation and does not take into account the addition of relativistic velocities, which will be covered in a later chapter. The predicted value of thrust is therefore slightly lower than that used in the final technical report on the experimental engine [1]. However, the low value for predicted thrust, even assuming that the rated power was actually obtained from the oven magnetrons used, illustrated the difficulties that would be faced in the programme of thrust measurements. The measurement problems were not helped by the total mass of the engine being above 15 kg and the total thermal output being around 1.2 kW. However, I was now committed to producing results, which I knew would come under rigorous review by real experts in their field. I was not wrong.

In the first set of thrust measurements, the experimental engine without the hermetic enclosure was mounted on a beam balance, with the majority of the module mass being counterbalanced, as shown in Figure 3.6.

The unbalanced mass was measured directly by a load cell connected to a digital meter via a low-drift amplifier. Access to the tuning plate adjustment and input tuning screw was at the top of the EMC enclosure. The engine could be mounted in a nominal configuration (thrust direction up) or an inverted configuration (thrust direction down). The magnetron was supplied with high-voltage DC power from a half-wave rectified power supply unit. The power was fed through three flexible links at the centre of the beam balance. The high cathode heater current (~10 A) necessitated multi-strand links with consequent mechanical damping of any beam movement. Also, the high current gave small but variable torques at the links when the magnetron was powered. Tests with a dummy load gave a mean force of +30 mg with a standard deviation of 29 mg. The load cell and amplifier contributed random noise to the results due to the high amplification required. Calibration tests gave a standard deviation of 20 mg with a calibration factor of 262 g/volt. For each test run, a test sequence of 50 seconds off, 50 seconds on and 50 seconds off was adopted, with load cell readings taken every 5 seconds.

Three test runs were carried out for each test configuration or tuning variation. Figure 3.7 shows the results for three consecutive runs in nominal and inverted configurations, together with three consecutive runs in nominal configuration but with the thruster deliberately detuned to give zero thrust. The horizontal scale in

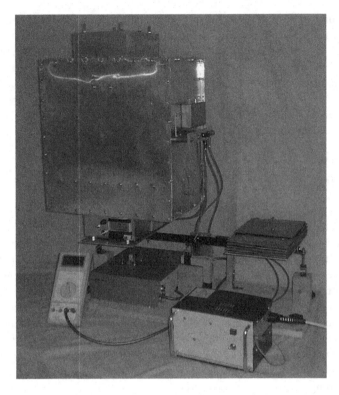

FIGURE 3.6 Experimental engine on beam balance.

Figure 3.7 is 50 seconds per division. Variable periods occurred between each run to allow the magnetron to cool. The zero thrust run illustrates the thermal buoyancy effect due to heated air outgassing from the EMC enclosure as the thruster module heats up, resulting in a decrease in total mass. The buoyancy effect can be calculated by adding the nominal and inverted test data and dividing by two. This is also plotted in Figure 3.7, and allowing for balance errors and run-to-run power variations, it is in good agreement with the zero thrust runs. The peak thrust, averaged over the three nominal and inverted runs, was 1.95 g. This appears to be in reasonable agreement with the predicted thrust. However, the peak is achieved after a 50-second run due to the time constant of the balance configuration. This was further investigated in later tests. First, however, it was necessary to address all the spurious forces that could be contributing to the results. A flurry of possible spurious forces was suggested by the reviewers of these initial results.

A total of 48 test runs were carried out in this series of tests, covering both nominal and inverted configurations over a range of input power and resonance tuning settings. Figure 3.8 shows the effect of the input tuning screw length on both input power, as measured on Det2, and thrust.

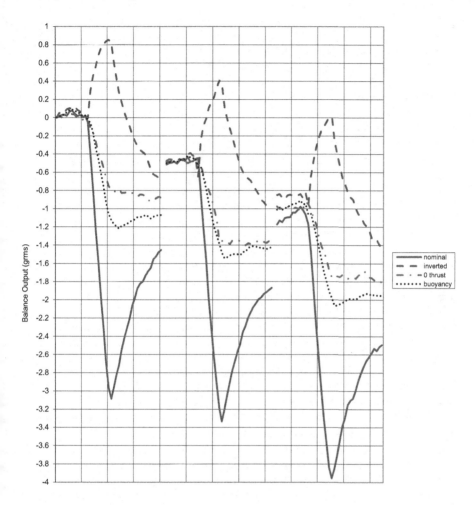

FIGURE 3.7 Initial test results for the experimental engine.

Similarly, Figure 3.9 shows input power and thrust measurements with variations in the position of the tuning plate. Note that it is possible to 'pull' the magnetron output frequency over a small tuning range, and evidence of the double peak of the main lobe of the output spectrum of this type of commercial magnetron is seen. The relationship between power and thrust is clearly demonstrated in these results, which were taken over separate test runs and power tests and included a change of magnetron. The wide bandwidth and pulling characteristics of the magnetron meant that the manufacturing tolerances of the cavity could be much larger than those that would be required for a narrow-band source. These early test results gave confidence that the experimental thruster was operating as expected.

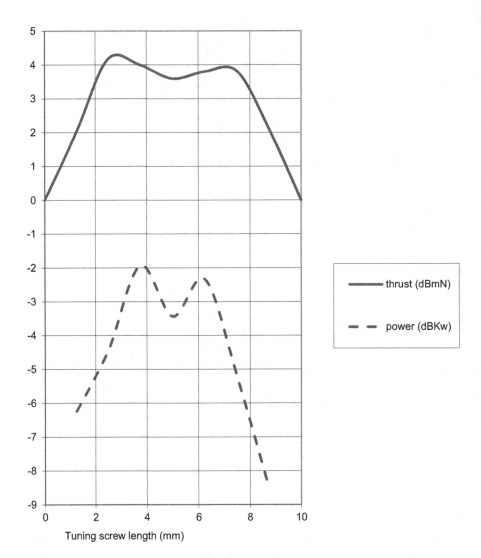

FIGURE 3.8   Effect of tuning screw length.

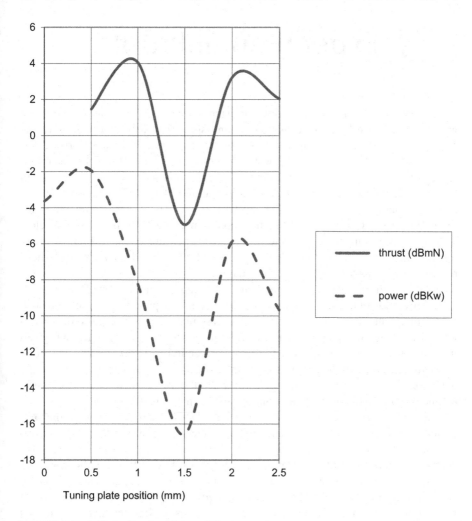

FIGURE 3.9 Effect of tuning plate position.

REFERENCE

1. Technical Report on the Experimental Microwave Thruster. Issue 2. September 2002. SMART Feasibility Study Ref 931. SBS South East. DTI.

4 Experimental Proof

Having established thruster operation consistent with design expectations, attention was directed towards the investigation of any spurious effects that may have the potential to produce similar test results. The effects considered were thermal, outgassing and electromagnetic. The test programme was carried out with constant settings for resonance and input tuning. The major thermal effect is the change in buoyancy due to the heating of the air within the unit. A total air volume of 0.03 m^3 and a cooling fan are employed. This enables 50 seconds of powered operation for three consecutive runs while limiting the maximum magnetron temperature to below 40°C. Actual air temperature rises were around 6°C per run, leading to thermal offset forces of around 1 g for the module test runs (see Figure 3.7). This agrees with theoretical calculations for the displaced air mass.

Following the initial test programme, the test setup was modified slightly to enable a rapid inversion of the unit to be achieved, and a dual test procedure was established. In this procedure, a single inverted run at a given input and resonance setting is followed by a nominal configuration run with the same settings. A typical result is given in Figure 4.1, which is comparable to the runs given in Figure 3.7 and illustrates that the sequence in which the test runs are carried out has no significance. Note also the deviations in slope during the early part of the 'on' periods due to the typical dip in magnetron power following switch-on, illustrated in Figure 4.2.

Thermal expansion effects could change the effective centre of gravity of the unit, giving rise to spurious torque values on the balance. To eliminate this effect, inversion of the unit was achieved by a "front to back" rotation. This rotation maintains symmetry in the vertical axis and, hence, the centre line of thrust. (The mounting points at the top and bottom of the unit are carefully set to avoid movement from the vertical following rotation.) Thus, spurious torques due to thermal expansion would be common to both nominal and inverted operations, and no test data differences between configurations can be attributed to thermal expansion effects.

To eliminate any spurious effects due to cooling fan operation, tests were carried out in a nominal configuration with both fans on and off. Figure 4.3 shows the results for a fan-on, fan-off, fan-on, sequence of test runs. It can be seen that rates of change of balance output during the heating phase (55 seconds to 100 seconds) and cooling phase (100 seconds to 150 seconds) are similar for all three runs, and it is clear that there are no direct effects on the test data due to cooling fan operation.

The buoyancy effect occurs because the unit is not hermetically sealed, and the heated air expands with some outgassing. At the end of the powered operation, the remaining higher-temperature air has a lower total mass. It is important that the outgassing process itself does not produce spurious forces in a specific direction. To test for this effect, openings in the enclosure and panel edges were modified with plastic tape. Whilst not achieving a total hermetic seal, the modification would certainly have changed any significant directionality in the outgassing process. Figure 4.4

DOI: 10.1201/9781003456759-4

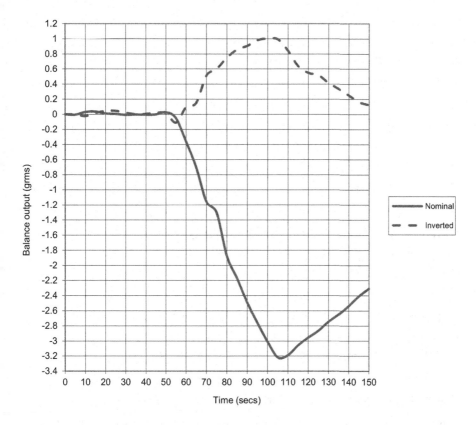

FIGURE 4.1 Results for dual test run.

compares the results of modified and unmodified runs. The rates of change in balance output are very similar for both runs. The peak negative output for the modified test is slightly less than for the unmodified test. This would be expected due to a lower total outgassed mass resulting from the restricted outgassing path. The conclusion can be drawn that there is no directional component to the outgassing process.

With the high DC currents being supplied to the magnetron cathode (~10 A) it was decided to eliminate any possibility of interaction with the Earth's magnetic field. Accordingly, tests were carried out at 0° orientation (approximately North/South) and at 90° orientation (approximately East/West). Figure 4.5 shows that within the normal spread of results, no significant effect can be detected.

From the 96 full power tests completed up to this point in the test programme, it became clear that all results exhibited a clear increase in measured force with time. The theory was addressed to see whether there was an explanation. This included a review of a possible quantum effect, which at the time was suggested as being the standard approach for physicists when they don't understand what is going on. A much simpler explanation was put forward, using classic physics to show that during the tests, an energy transfer from the stored energy in the cavity to the potential energy in the balance was taking place. It was then realised that the actual power input from the magnetron comprised 50 Hz pulses due to the

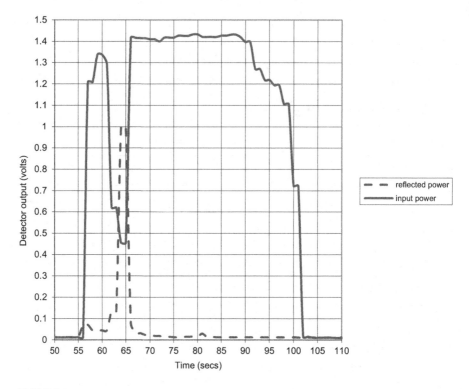

FIGURE 4.2   Input and reflected power from switch-on.

half-wave rectified DC input to the magnetron. This provided a method of demonstrating that not only was the measured force not due to spurious forces, as shown by the previous test data, but could only be due to EmDrive thrust. The theory of energy exchange following switch-on was to prove the solution to an in-orbit problem in a commercial experimental satellite many years later. However, for the time being, it remained a theory that needed to be proved, and once proved, it remained out of the public domain until the final report on the feasibility study [1] was released by SPR Ltd. in 2016.

The first task in this proof was to minimise the buoyancy effect and carry out a series of tests using two precision balances and the hermetic enclosure. In balance configuration 1, the thruster was mounted on a beam balance with 15.544 kg of the thruster mass offset by a counterweight on the opposite side of the beam. The mass difference was measured on a 110 g electronic balance with a resolution of 1 mg and a spring constant of 1.635×10^3 N/m. In balance configuration 2, the engine was mounted directly on a 16 kg electronic balance with a resolution of 100 mg and a spring constant of 1.825×10^5 N/m. Figure 4.6 shows the engine, complete with hermetic enclosure, mounted in balance configuration 2. A draught shield was built to fully enclose the engine and balance, with high-voltage and data connections in the side wall. The object of the programme was to obtain results from a hermetically sealed thruster and to investigate the thrust profile seen during the test runs using the load cell.

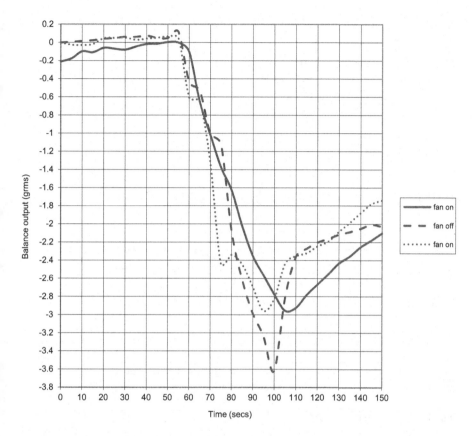

FIGURE 4.3 Cooling fan tests.

Maintaining a hermetic seal using the enclosure over a full test programme proved difficult. However, two sets of tests carried out in balance configuration 2 illustrate the results of minimising the buoyancy effect by significantly decreasing the outgassing of heated air. Figure 4.7 gives results without the hermetic enclosure for nominal and inverted thruster configurations. Also given is a zero thrust result with the thruster deliberately de-tuned. Note that although the buoyancy effect gives a negative force of 1.4 g, the positive and negative thrusts are only 0.2 g.

Figure 4.8 gives results for the thruster inside the hermetic enclosure. For these tests, a positive thrust of 0.4 g after 20 seconds of power and a negative thrust of 0.6 g after 30 seconds are achieved. A negative buoyancy force of 0.1 g is identifiable for the detuned test run. Figures 4.7 and 4.8 also clearly illustrate the much lower rate of thrust increase measured in balance configuration 2 compared to those measured in the load cell tests. This is completely in accordance with the law of conservation of momentum, where the thrust and reaction forces should completely cancel each other out when there is no acceleration. The fact that any thrust increase is measured at all is due to the spring constant of the balance causing a tiny acceleration to take place.

In contrast, Figure 4.9, which gives results in balance configuration 1, shows higher thrust rates and peak thrust values comparable to those obtained using the

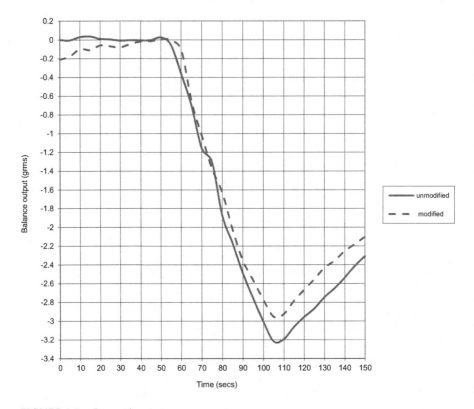

FIGURE 4.4 Outgassing tests.

load cell. Figure 4.9 also illustrates the very reduced buoyancy effect using the hermetic enclosure. This is calculated as the difference between the inverted and nominal test results.

Clearly, there was a need to explain this set of results by investigating the theory of the energy exchange between the stored electromagnetic energy in the cavity and the potential energy stored in the spring within the balance at the end of the powered run.

The equation for the thrust output of the microwave thruster is given by equation (3.4).

$$T = \frac{2PQu}{c}\left\{\frac{\lambda_0}{\lambda_{g_1}} - \frac{\lambda d}{\lambda_{g_2}\sqrt{\varepsilon R}}\right\}$$

Let D_f be the thruster design factor where:

$$D_f = Qu\left\{\frac{\lambda_0}{\lambda_{g_1}} - \frac{\lambda d}{\lambda_{g_2}\sqrt{\varepsilon R}}\right\}$$

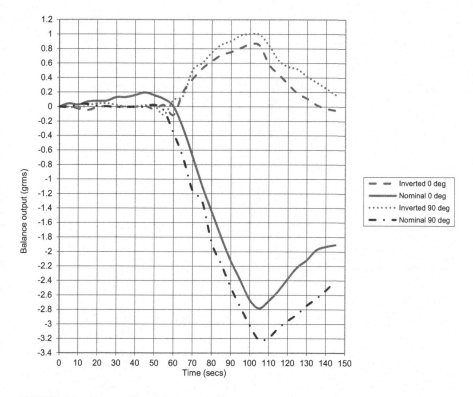

FIGURE 4.5   Tests at 0° and 90° orientations.

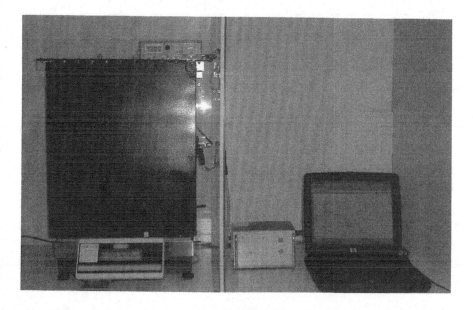

FIGURE 4.6   Experimental engine in balance configuration 2.

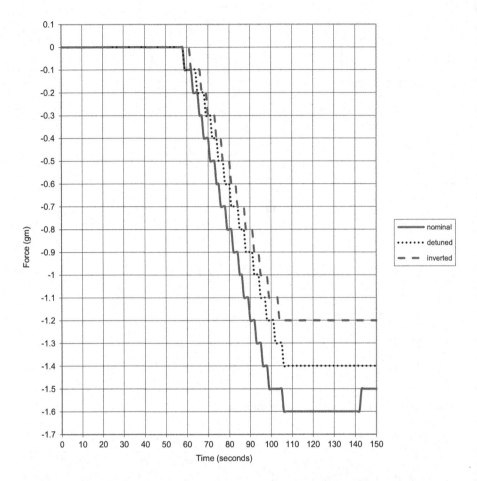

FIGURE 4.7 Balance configuration 2 tests without hermetic enclosure.

Then

$$T = \frac{2PDf}{c} \tag{4.1}$$

Now the thrust is a result of the rate of change of momentum of the electromagnetic (EM) wave within the thruster.

Let R_e be the change in momentum of the EM wave over time t_m

Then, from the basic equation force = mass x acceleration:

$$T = \frac{R_e}{t_m}$$

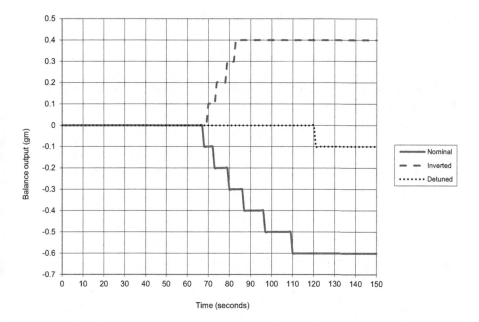

FIGURE 4.8 Balance configuration 2 tests with hermetic enclosure.

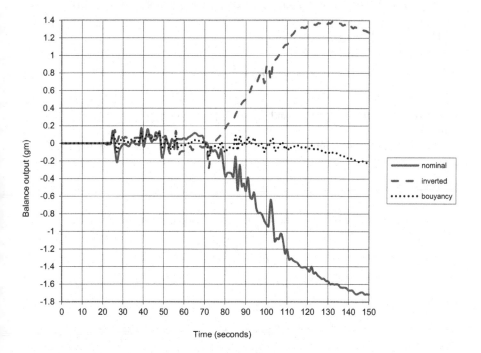

FIGURE 4.9 Balance configuration 1 tests with hermetic enclosure.

thus

$$R_e = \frac{t_m 2PDf}{c} \qquad (4.2)$$

For an unrestrained thruster, the momentum transferred from the EM wave to the thruster is given by:

$$R_m = MV \qquad (4.3)$$

where M is the mass of the thruster and V is the velocity attained in time t_m.

Then, by the conservation of momentum,

$$MV = \frac{t_m 2PDf}{c}$$

Thus, the time to raise the velocity of the thruster by V is:

$$t_m = \frac{MVc}{2PDf} \qquad (4.4)$$

However, if the thruster is restrained by a spring balance, the kinetic energy of the moving thruster is transferred to the potential energy stored in the spring as the spring is compressed.

This potential energy is given by:

$$E_p = \frac{Kl^2}{2} \qquad (4.5)$$

where K is the spring constant and l is the spring compression due to the energy transfer.

Now the apparent change in mass m displayed on the balance is the measured thrust T_m.

Thus $m = \dfrac{T_m}{g}$

The kinetic energy stored in the spring at any time is therefore given by:

$$E_k = \frac{T_m V^2}{2g} \qquad (4.6)$$

By conservation of energy

$$\frac{T_m V^2}{2g} = \frac{Kl^2}{2}$$

Thus,

$$V = l\sqrt{\frac{Kg}{T_m}} \qquad (4.7)$$

Now the time required to raise the velocity of the thruster by V is given in equation (4.4).

From equations (4.4) and (4.5):

$$t_m = \frac{Mcl}{2PDf}\sqrt{\frac{Kg}{T_m}} \qquad (4.8)$$

The experimental thruster is powered by a magnetron supplied by a 50 Hz half-wave rectified voltage. The power of the EM wave within the thruster is therefore a pulse waveform with 10 ms pulses of $\sin^2 x$ shape at a repetition rate of 50 pulses per second. This is illustrated in the oscilloscope trace of the output from the power detector mounted within the cavity (Det 2), given in Figure 4.10.

This pulsed thrust produces a balanced response that modifies each pulse by the time required to achieve energy transfer (t_m ms) and by the mechanical time constant of the balance itself.

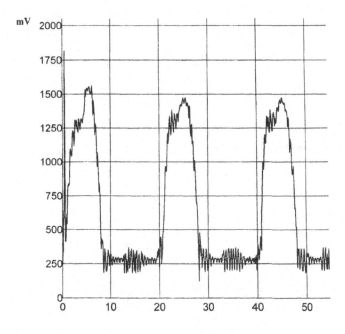

FIGURE 4.10   Oscilloscope trace of power from Det. 2.

Let the peak thrust in each pulse be T_p and the damped balance time constant be t_b ms.

Then, in the 10 ms pulse period from $t = 0$ to $t = 10$,

$$T = T_p \left(\frac{t}{t_m} \right) \left(1 - e^{\frac{-t}{t_b}} \right) \sin^2 \pi \frac{t}{10} \qquad (4.9)$$

During the inter-pulse period from $t = 10$ ms to $t = 20$ ms

$$T = T_{10} e^{-\left(\frac{t-10}{t_b} \right)} \qquad (4.10)$$

where T_{10} is the measured value of thrust at 10 ms.

After 20 ms, the cycle repeats, with T for the next cycle being added to the residual value of T from the first cycle.

The solution to equations (4.9) and (4.10) is given in Figure 4.11 for the first two cycles of a test run in balance configuration 1.

Thus, for a test run of hundreds of cycles, the measurement of thrust will increment towards the peak thrust T_p, with the rate of increment dependent on the mass of the thruster and the spring constant.

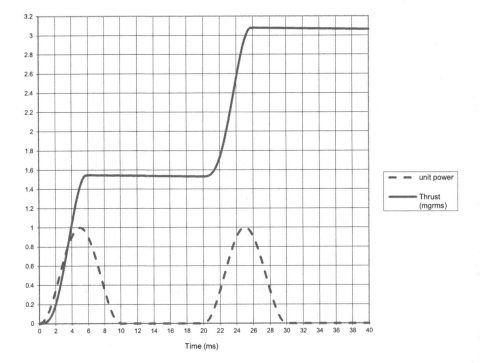

FIGURE 4.11 Power and thrust over two cycles.

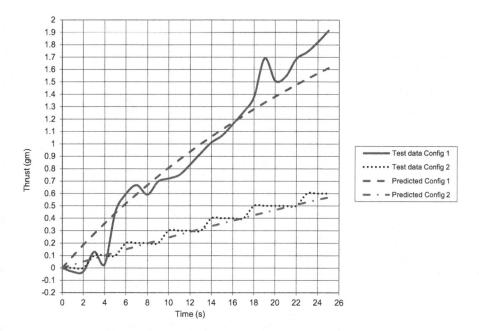

FIGURE 4.12   Predicted and experimental thrust.

For the experimental thruster, the root mean squared (RMS) power input (P) is 850 W, the design factor (D_f) is 2470, and the mass (M) is 15.59 kg. From equations (4.9) and (4.10), the thrust profiles for balance configurations 1 and 2 were calculated. In Figure 4.12, these predictions are compared to the thrust data taken over 25 seconds of test runs in balance configurations 1 and 2.

In a normal test run, the power is on for 50 seconds. In the first 5 to 15 seconds, the cathode is heating up prior to microwave power being produced. Over the next 10 seconds, the rapid temperature rise of the magnetron itself causes frequency instability, giving rise to the variation in output power illustrated in Figure 4.2. At the end of the 50-second period, the magnetron is reaching its maximum operating temperature, causing further frequency instability. There are therefore typically 25 seconds of stable thruster operation during a test run. Figure 4.12 shows that over this 25-seconds period, there is close agreement between the predicted and measured results in both balance configuration 1 and configuration 2.

It is now worth considering the results if the force measured on the balance had been due to a constantly applied force (e.g., a mass change). It would be expected that the balance response would merely follow the balance time constants. Before test runs, the balances are calibrated by adding precision weights. The scaled calibration responses are given in Figures 4.13 and 4.14, together with the predicted responses using the balance time constants. Allowing for the balance oscillations that occur experimentally, the predicted and calibrated responses agree closely.

Similarly, if the force measured on the balance had been due to a linear change in the mass (e.g., due to thermal effects), the responses would be the linear change modified by the balance time constants illustrated in Figure 4.15.

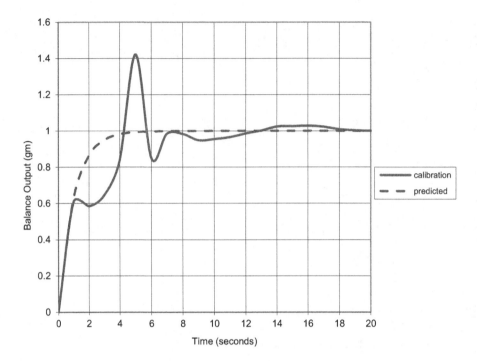

FIGURE 4.13 Balance configuration 1 response to a constant spurious force.

Inspection of Figures 4.13–4.15 shows that in each case of an applied force due to a spurious effect, the predicted responses of the two balance configurations are quite different from the response due to the thruster output of equation (1.5). Specifically, for each spurious force, the rate of increase in predicted force is higher for configuration 2 than for configuration 1. However, for the measured engine output given in Figure 4.12, the rate of increase in thrust is higher for configuration 1 than for configuration 2. Clearly, the test results obtained from the experimental engine are not due to the spurious forces considered.

At this point in the programme, having accepted that the results were not due to the spurious forces proposed by the reviewers, the question was asked whether they were due to spurious forces that nobody had thought of. This was a challenge that was addressed by considering what the balance responses would be for a pulsed force waveform rather than for the pulsed momentum produced by the thruster. Such a force waveform could occur if there were a spurious pulsed electromotive force present rather than the thruster output, as given by equation (1.5).

Let this peak force be equal to the apparent change in mass m times the gravitational acceleration g.

Then the potential energy stored in the spring balance for an applied force is given by

$$E_p = mgl \tag{4.11}$$

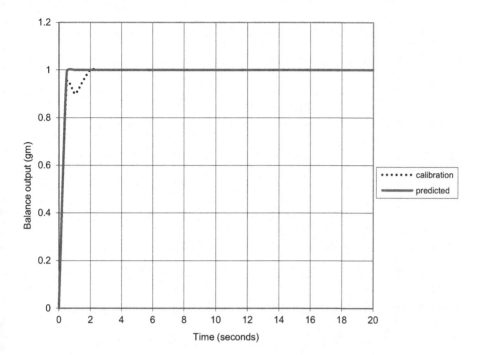

FIGURE 4.14 Balance configuration 2 response to a constant spurious force.

Where l is the spring compression,
    Thus, by conservation of energy

$$mgl = \frac{Kl^2}{2} \tag{4.12}$$

Now compression l is given by $l = Vt_f$ where t_f is the time for the force to achieve compression l, and velocity V is given by $V = gt_f$
    Then substituting in equation (4.12) gives

$$t_f = \sqrt{\frac{2m}{K}} \tag{4.13}$$

Applying t_f to equations (4.9) and (4.10) in place of t_m gives predicted responses for the two balance configurations and is shown in Figure 4.16.
    Once again, the rate of increase in the predicted force is higher for configuration 2 than it is for configuration 1, which is opposite to that measured for the engine and given in Figure 4.12. Pulsed spurious forces were therefore ruled out.
    As a final verification of the thrust profile theory, the pulsed thrust was directly measured on an oscilloscope. The original beam balance and load cell, as shown in

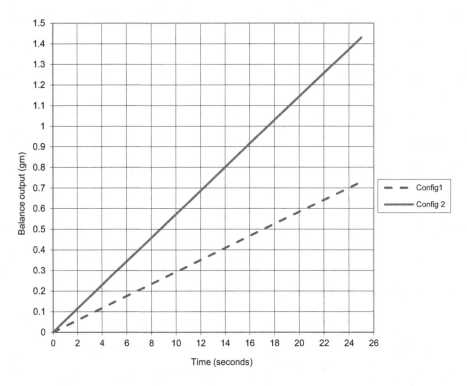

FIGURE 4.15 Predicted response for a linear spurious force.

Figure 3.6, were used. The load cell electronics were modified to include a two-stage amplifier of gain 60×10^3 with AC coupling and low-pass filtering. The measured positive and negative responses of this balance configuration are given in Figure 4.17, together with the predicted responses using the theoretical balance time constant.

The very high overall gain from load cell to oscilloscope and the proximity of the high voltages (5 kV) caused considerable electromagnetic compatability (EMC) problems, particularly as the wanted data (thrust pulses) and the voltage transients were both at 50 Hz. Three sets of measurements were taken, one with the thruster in the nominal configuration and one in the inverted configuration. A further set was taken with additional counterbalance mass so that the load cell was unloaded. This third test configuration gave a reference EMC response, and test data was processed by subtracting the EMC response from the thrust measurements. As the measurements were from ac-coupled amplifier stages, the results were then integrated over two cycles to give the thrust output shown in Figure 4.18.

Also given in Figure 4.18 are the inverted and nominal thrust profiles predicted for the load cell balance configuration using equations (4.9) and (4.10). There is a clear correlation between the predicted and measured thrust data over two cycles for both nominal and inverted configurations.

The feasibility project was reported to DTI in a full technical report [1], which was reviewed by technical experts within DTI and MoD. David Fearn, the UK's leading expert in electrical propulsion at the Space Department at the Royal Aircraft

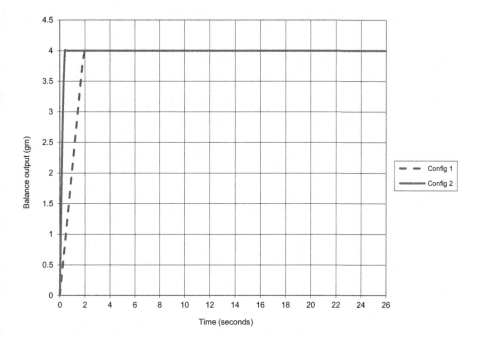

FIGURE 4.16 Predicted balance responses for spurious pulsed forces.

Establishment (RAE) Farnborough, was included in the review process. John Spiller, a space systems consultant with a long career in design and analysis for many European space programmes, prepared a separate independent report [2]. Following this lengthy review process, the final report contained the following list of agreed-upon conclusions:

Conclusions from the Experimental Engine Technical Report to The Department of Trade and Industry (DTI).

1. A theory has been developed for a propulsion technique that allows direct conversion from electrical energy at microwave frequencies to thrust.
2. An expression has been developed to enable the thrust from such a system to be calculated.
3. An experimental thruster was designed using a mathematical model to establish the basic dimensions and electrical characteristics.
4. An experimental 850 W thruster operating in the S band was manufactured, and resonant operation at microwave frequencies was tested using low power sources.
5. The tuning dimensions for resonance at the nominal operating frequency of 2541 MHz and at the lower frequency limit of 2432 MHz were measured. These dimensions agreed closely with the tuning positions predicted by the model.
6. A further test of short-circuiting the dielectric section resulted in no resonance being detected, proving that propagation was taking place as designed.

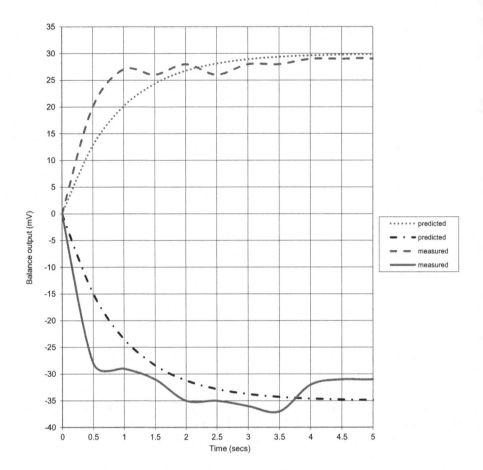

FIGURE 4.17   Predicted and measured responses for beam balance and load cell.

7. The experimental thruster was successfully powered by an 850 W magnetron approximately 450 times over periods of up to 50 seconds. A full range of thrust, power and temperature measurements were recorded. During the programme, five different magnetrons were used.
8. The test programme included tests in both a nominal and an inverted position on three different balances. In all tests, when correctly tuned, the thruster gave an upward force in the nominal position and a downward force in the inverted position, in accordance with the design.
9. Peak thrusts measured were comparable to those predicted from the theoretical expression.
10. When not correctly tuned, the thruster gives reduced thrust or no measurable thrust. Results over a range of input and resonance tuning adjustments showed a good correlation between power and thrust measurements.
11. The sequence in which test runs were carried out was shown to have no influence on the thrust measured.

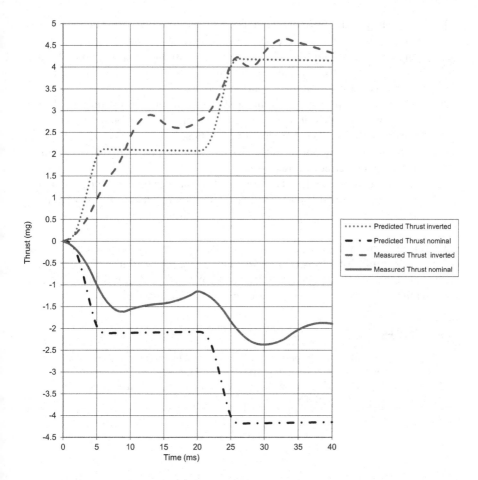

FIGURE 4.18 Predicted and measured thrust over two cycles.

12. The operation of the cooling fan was shown to have no influence on the thrust measured.

13. The direction of the outgassing of cooling air from the thruster was shown to have no influence on the thrust measured.

14. The orientation with respect to the Earth's magnetic field was shown to have no influence on the thrust measured.

15. Operation within a hermetic enclosure clearly demonstrated the reduction of the buoyancy offset due to the outgassing of cooling air.

16. Expressions were derived for the thrust/time profile to predict the pulsed thrust output due to a half-wave rectified voltage input to the magnetron.

17. These expressions were used to predict the thrust/time profiles expected when the thruster was tested on two balances with widely different spring constants.

18. The thrust/time profiles measured agreed closely with those predicted for both balances.

19. Force/time profiles were predicted for assumed spurious forces due to mass change, thermal slopes and electromagnetic pulse forces. In each case, the change in profile with the different balance configurations was opposite to the profile change actually measured. This finding totally confirms the earlier tests that eliminated these spurious forces.
20. The thrust/time profile equations predicted a stepped thrust output when viewed over a number of pulse cycles. This stepped thrust output was measured using the processed output of an oscilloscope.
21. The pulsed thrust output correlated closely with the predicted output in both nominal and inverted configurations.
22. It is concluded that the test data presented verifies the theory of operation of the microwave thruster, and thus, for the first time, a method of propulsion that does not rely on propellant has been demonstrated.

Following this in-depth peer review process, the conclusions were completely supportive of the EmDrive concept. To further emphasise their enthusiasm, DTI invited us to apply for more funding via an R&D grant. When the UK government offers to give you more money, you know you are on to something special.

REFERENCES

1. Technical Report on the Experimental Microwave Thruster. Issue 2. September 2002. SMART Feasibility Study Ref 931. SBS South East. DTI.
2. Spiller J W. Review of Technical Report on the Experimental Microwave Thruster. 29 October 2002. JWS-SPR-TN-005.

5 Demonstrator Engine

The experimental thruster, tested during the feasibility study, clearly demonstrated the principles of operation and confirmed the theoretical performance predictions. However, thermal constraints limited the test runs to less than 50 seconds; the commercial magnetron gave a pulsed thrust output, and thrust was measured on a series of static balance rigs. The performance of the experimental thruster was also limited by the relatively low Q operation that could be achieved due to the use of a dielectric section.

Thus, although a detailed review of the technical results confirmed the validity of the concept, further large-scale support of the work would need a more convincing demonstration. The purpose of the demonstrator engine was to provide a clear visual demonstration of operation as well as the means of carrying out a detailed development programme. It was proposed that the demonstrator engine be tested on a rotary test rig consisting of a turntable mounted on an air bearing. The torque generated by the engine would enable rotary acceleration to be demonstrated and measured. The demonstrator engine would incorporate a water-cooled magnetron and radiator, together with a continuous high-voltage power supply. The resulting continuous thrust would enable rotation to build up and provide clear visual proof of operation. By mounting the engine in three 90° configurations, forward, reverse and zero-torque operation would be demonstrated. Theoretical work had indicated that an increase in output thrust of almost ten times that obtained with the experimental thruster should be available. Due to the nominal fixed frequency operation of a magnetron, it was necessary to design the cavity to retain the same axial length whilst undergoing a significant temperature rise, as over 1 kW of power would be dissipated during continuous operation.

The proposal for the demonstrator engine project was submitted to DTI in July 2003 and included an initial assessment of the commercial return from the application of EmDrive to the launch of typical communication satellites into geostationary orbit (GEO). The EmDrive engines would be used in the later stages of satellite launches (GEO transfer) and for maintaining operational orbit. The performance of EmDrive is such that the launch mass of the satellite would be halved and its lifetime significantly extended. The use of ion propulsion has demonstrated that electric propulsion techniques are viable in certain mission applications; however, Table 5.1 was presented to illustrate the considerable cost and performance benefits that EmDrive propulsion would offer to commercial communication satellite operators.

Current propulsion was assumed to be a bipropellant apogee engine and attitude and orbit control system (AOCS) thrusters, whereas the ion propulsion data was based on the latest European Space Agency (ESA) thruster that was later used in their SMART-1 mission.In this case SMART stood for Small Missions for Advanced Research in Technology. The EmDrive system assumed an S-band primary engine. The GEO transfer for current systems assumed a conventional elliptic transfer orbit,

DOI: 10.1201/9781003456759-5

TABLE 5.1

2003 Estimate of Launch Costs

	Current	Ion	EmDrive	
Comms payload mass	345	345	345	kg
Satellite launch mass	3040	1990	1312	kg
Satellite DC Power	6	6	6	kW
Height	2.8	2.8	2.8	m
Length	1.7	1.7	1.7	m
Width	2.5	2.5	1.3	m
GEO transfer time	<1	312	36	days
Propulsion system cost	3.1	41.7	3.9	£million
Total satellite cost	57.9	96.5	56.4	£million
Launch costs	79.8	52.2	34.4	£million
Total cost	137.7	148.7	90.8	£million

whereas the ion and EmDrive propulsion systems assumed a spiral (Edelbaum) transfer from a low Earth orbit. The launch capability and costs were scaled from real space industry costs available at the time and combined with the latest 10-year launch prediction issued by the well-respected Teal Group. For GEO satellites alone, the predicted 373 launches would save an estimated £15.5 billion. This would be the driver for establishing a manufacturing business worth £3.8 billion over 10 years.

To support the proposal, a copy of the Patent Office search report for our second patent, GB2334761 [1], was included. This stated that the work was "at the leading edge of innovation" when the patent was granted in April 2000. In September 2003, a grant of £81,291 was offered for the Research and Development Project Reference 1939 by the South East England Development Agency (SEEDA), a regional department of DTI. This was gratefully taken up, and we launched into 3 years of exciting work with the eventual aim of demonstrating that EmDrive could accelerate a total mass of around 100 kg in a simulated zero-gravity environment. It would be seen to be moving!

Figure 5.1 gives a simplified schematic diagram of the demonstrator engine and its supporting units. The basic thruster consists of a cavity resonator and a resonance tuner. The thruster is fed microwave power from the magnetron via a feed assembly comprising a circulator and load (which together form an isolator), a dual coupler, and an input tuner. The magnetron and load are cooled by a pumped coolant circuit containing three radiators and a preheat unit. The magnetron is powered by a high-voltage power supply unit (HVPSU), and the engine is controlled and monitored by a telemetry and command unit (TCU). This unit, together with the HVPSU and a laptop telemetry computer is mounted separately on an instrumentation platform on the test rig. A main power unit and remote command unit are mounted off the test rig.

The magnetron, feed assembly and thruster are mounted in an open steel thrust frame, which provides attachment to test rigs in five thrust configurations. These give the following thrust vector directions:

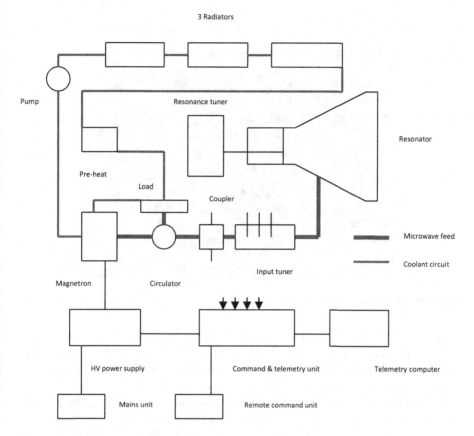

FIGURE 5.1 Demonstrator engine schematic.

Vertical up
Vertical down
Horizontal forward
Horizontal reverse
Horizontal axial

The thermal subsystem, complete with telemetry and power interfaces, is mounted in an open aluminium U frame, which fits over and is attached to the thrust frame. The complete demonstrator engine is shown in Figure 5.2.

The software developed during the design of the experimental thruster was updated and used, in an iterative design process, to optimise the geometry of the resonator and tuning sections of the demonstrator engine. The resulting design (at the seventh iteration) was the basis for further consideration of the thermal stability and tuning resolution requirements. A review of the potential for resonance in spurious modes was carried out by running the software for twelve different modes, in both Transverse Electric (TE) and Transverse Magnetic (TM) modes. These modes

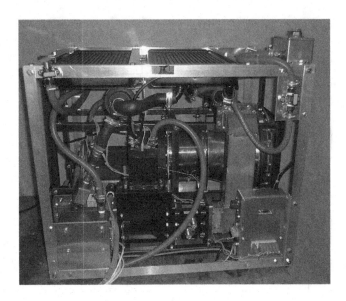

FIGURE 5.2   Demonstrator engine.

were from TE01 to TE22 and from TM01 to TM22. The design resulted in a maximum diameter of 280 mm operating at 2,450 MHz with a design factor of 0.844. This can be compared to the earlier experimental thruster with a maximum diameter of 160 mm and a design factor of 0.497 at the same operating frequency.

A signal level analysis was carried out for the resonator coupled to the feed and instrumentation components to allow specifications for these items to be prepared. To support the microwave design work, a low-power test bench was set up. This was calibrated against the experimental thruster and then used to verify the microwave design of the demonstrator engine. The design of the tuning section incorporated a stepper motor, gear train and lead screw, together with the necessary electronics, to give the very high-resolution tuning required by the Q value of 72,000 that was theoretically attainable with the resonator.

The mechanical design was based on a fabricated copper resonator and tuner supported by brass plates held together with a combination of brass and invar tie rods. This construction, together with a steel lead screw, allows the different expansion coefficients of the four materials to be used to provide inherent thermal compensation. This was considered essential to maintaining a stable resonance frequency over the required temperature range. Thermal design software was developed to enable optimisation of the dimensions to obtain full thermal compensation over a 35° temperature range. The thruster is illustrated in Figure 5.3.

Figure 5.2 shows that the WR284 waveguide output of the magnetron, mounted at the bottom left of the thrust frame, is connected through a 90° E bend assembly to an isolator. The isolator consists of a circulator and a load. The circulator, type WR284CIRC3A, is rated at 3 kW with a 0.2 dB insertion loss. The water-cooled load, type GA1201, is also rated at 3kW with a minimum return loss of 23 dB. The load is specified up to a maximum input water temperature of 50°C. Power is then fed

FIGURE 5.3 Demonstrator thruster.

through a dual coupler, type GA3102, to monitor forward and reflected power. The coupling factor is 56 dB, with a minimum specified directivity of 25 dB. The feed from the coupler is via a waveguide assembly containing both a 90° H bend and a 90° E bend to reach the input tuner, situated above the large diameter end of the resonator. This position enables easy access to the stub adjusters. The GA1001 precision 3-stub tuner has stubs spaced at ¼ wavelength intervals, offset 1/16 guide wavelength from the centre. The stub housings are designed with reactive chokes for high-Q applications. Power is finally fed to the resonator input via a U-shaped waveguide assembly containing two 90° E bends. All waveguides are standard WR284 (WG10) brass with type CPR 284F brass flanges. Although the resulting feed assembly is heavy compared to a flight waveguide, it is rugged and easily fabricated.

The magnetron and power supply were procured as a complete microwave generator set, type GA4305, from a US supplier. The set consisted of a separate magnetron head unit and a switched-mode power supply with an interconnecting cable harness. The magnetron itself is a water-cooled 2M137 device rated at 1.2 kW at an operating frequency of 2,450 MHz ± 30 MHz. It is mounted in the head unit, which also includes the filament transformer, control interlocks and feedback electronics for the power supply. Maximum power is specified for an input water temperature of 35°C, with a nominal lifetime of 3,000 hours under a matched load. The power supply provides constant anode current at a cathode voltage of −4.5 kV. The filament transformer is supplied at 230 V AC. The microwave power level is controlled by an analogue signal from the TCU. Alarm and trip circuits are provided, with all power and enable circuits controlled via the TCU.

The thermal conditions within the engine are quite different for narrow-band or broadband magnetron operation. For broadband operation, when the magnetron is free-running and not locked to the cavity resonant frequency, the majority of the input microwave power is reflected at the cavity input and dissipated in the isolator. To provide the necessary cooling, water is pumped through both the magnetron and the isolator load. For narrowband operation, when the magnetron output frequency pulls in and locks to the cavity resonant frequency, the microwave power is largely dissipated in the cavity itself. To support the design of the water cooling and radiating thermal subsystem, a test rig was manufactured and a series of thermal tests were carried out. The test rig was constructed to similar dimensions and with the same layout as the engine and was capable of being mounted in the same manner. The coolant loop incorporated a 2 kW pre-heat unit and pump, together with two radiators. Mounting the pump between the two radiators ensured the pump remained primed in all three mounting positions. The tests were carried out using eight thermocouples linked to a computer via data logging electronics and enabled design verification of all the subsystem components. An intercooler unit was manufactured using fabrication techniques proposed for the engine. This was incorporated into the final thermal tests and confirmed the thermal radiation calculations. The preheat unit was retained in the actual engine coolant loop to enable thermal simulation tests to be carried out during the thrust measurement programme. The unit also provided a method of preheating the coolant to limit the initial temperature rise of the magnetron. The third radiator was incorporated into the subsystem to enable long test runs to be achieved during dynamic testing.

The design of the telemetry and command subsystems, together with power supply considerations, resulted in the engine being connected to four major units. A HVPSU provided the magnetron cathode and filament currents. A TCU provided power and control commands for the magnetron, coolant pump, preheat unit and tuning drive. The TCU also interfaced and processed up to 8 digital and 11 analogue telemetry channels. These were used to monitor temperatures, microwave power measurements, tuner positional data and engine component status. The TCU also provided displays for temperature, tuning, and power readout. During test runs, a Remote Command Unit was used for control purposes, and a Telemetry Computer was used to record all engine telemetry and test instrumentation data.

The work on our experimental engine and the award of the demonstrator engine grant were reported at our first public presentation at the BIS Symposium in Brighton in October 2004. The paper entitled *The Development of a Microwave Engine for Spacecraft Propulsion* was published in the BIS journal [2], edited by Dave Fearn, who had been part of the DTI review team for the Experimental Engine technical report.

As with the experimental engine, once the demonstrator engine cavity had been manufactured, an extended period of alignment, tuning and small signal testing was carried out. During this period, an internal-shaped large end plate, based on the geometry shown in Figure 2.7, was carefully aligned with the small end plate, which forms the moving part of the cavity tuner. Alignment was continued throughout this period with the object of always aiming for the highest cavity Q value. To enable sweep tests of the resonance tuning mechanism to be carried out, breadboard circuits to drive the stepper motor and monitor tuner position and resonator power were designed and built. These circuits were then incorporated into the TCU. The small signal test bench is shown in Figure 5.4, where the cavity is mounted in the thrust frame.

The first series of tests enabled the thruster to be characterised over its full tuning range. The position of the tuning plate was stepped through its range, while the output from a detector probe mounted in the resonator was recorded on

FIGURE 5.4   Small signal test bench.

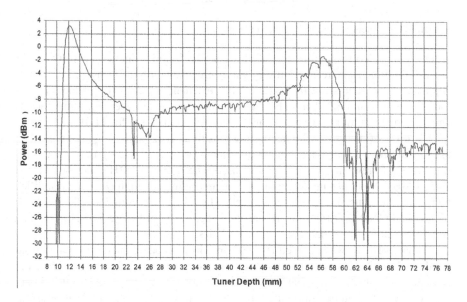

FIGURE 5.5   Detector output for full cavity tuner sweep.

the telemetry computer. A crystal-stabilised low-power signal at a frequency of 2,449.9 MHz was fed into a loop mounted at the input port of the cavity.

Figure 5.5 gives the detector output for the full tuner depth range of 8 mm to 78 mm. As the input impedance was unmatched, a drop in the detected signal was seen as the tuner went through resonance, with a single clear resonance peak detected for both TE01 and TE11 modes. The tuner depths at which resonance was detected, i.e., 10.1 mm for TE01 and 64.5 mm for TE11, fall within the range of theoretical depths that were obtained by running the design software for the thruster dimensions, plus and minus the manufacturing tolerances. A small resonant peak is seen

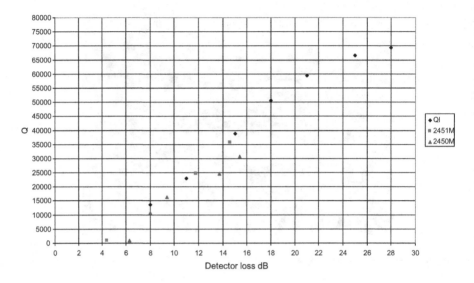

FIGURE 5.6   Theoretical and measured Q values.

around 26 mm, corresponding to the TM01 mode. Running the design software for the as-built dimension, resulted in very good agreement with the measured resonance depth for the operational TE01 mode. The two resonance peaks give Q values of 8,194 for TE01 mode and 5,257 for TE11 mode. This is in accordance with field theory, which predicts a lower surface resistance and hence a higher Q for TE01 mode.

A number of detectors with decreasing sensitivity (i.e., increasing loss) were then used to produce a more accurate measurement of Q at the TE01 resonance position. Two different signal sources were used, one at 2,449.9 MHz and the other at 2,451.1 MHz. Note that any direct measurement of this very high Q value is modified by the loss introduced by the detector itself. The results are plotted together with the theoretically loaded Q (Q_L) for increasing detector loss in Figure 5.6. The test data points fit well with the theoretical characteristics.

A series of tests were carried out with the thrust module enclosed in a thermal hood, enabling a 32° ambient temperature rise to be achieved. The results are given in Figure 5.7, which shows that the resonant frequency remains within +160 kHz to −70 kHz over this temperature range. This was well within the required range of stability. A total of 57 small signal sweep tests were carried out, producing a set of data that validated the microwave and thermal compensation design processes.

The initial high-power test runs were carried out with the input tuner settings that had been optimised from the small-signal tests. A sequence of measurements was then carried out, with increasing magnetron power settings, while the temperature telemetry was carefully monitored. The initial two-radiator configuration allowed a few consecutive 60-second test runs to be carried out without excessive coolant temperature rise. (Note that the 60 seconds refers to the magnetron's 'on' period.). However, later in the vertical test programme, coolant flow problems occurred in the vertical down attitude, leading to one case of a magnetron over-temperature trip and the eventual failure of the preheat element. In the final three radiator configurations,

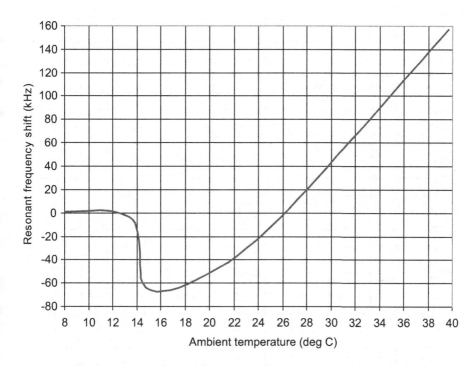

FIGURE 5.7 Thermal stability test.

horizontal test runs of 120 seconds duration produced a magnetron temperature rise of 4.3°C, and an isolator temperature increase of 11.62°C, for a coolant temperature increase of only 0.43°C.

The coolant is monitored at the magnetron input, and the high thermal capacity of the whole loop causes the temperature to continue to rise as the preheat isolator and magnetron sensors record initial cooling after switching off. Eventually, these four temperatures equalise during the slow cool-down period. The preheat temperature is the coolant temperature in the pre-heat unit and records the highest temperature reached in the loop. The overall temperature increases for the tuner section (0.35°C), the resonator section (0.4°C) and the input tuner (0.8°C) give an indication of the microwave loss distribution in the thruster. With the magnetron at a mid-power setting, the engine was then tested over the full resonator tuning range. Figure 5.8 gives a composite result of the reflected power for a number of swept tuner tests. These tests showed TE01 resonance over an approximately 5 MHz bandwidth, which corresponds with the quoted output bandwidth of the magnetron. A TE11 resonance mode was also measured. Both resonance points aligned closely with those obtained during the low-power tests and with the theoretical design.

The calibration of the microwave power telemetry was carried out by first calibrating the DC circuits for a reference input voltage. The DC measurement could then be referred to the microwave input using the calibration data provided by the manufacturer of the power meter units. The input to these units from the coupler was then calibrated by connecting the coupler between the magnetron output and

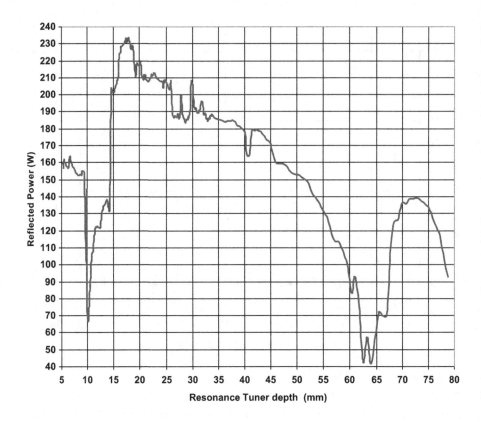

FIGURE 5.8   Full resonance tuner sweep.

the circulator input. This enabled calibration using only forward power flow. The calibration was completed by connecting the coupler between the circulator output and a waveguide short circuit. This gave a final calibration for simultaneous, equal, forward and reflected power flows. The Demonstration Engine was now ready for its programme of static thrust and dynamic acceleration tests, and thus the first attempt to demonstrate that EmDrive does indeed satisfy Newton's laws.

REFERENCES

1. Shawyer R. *Microwave Thruster for Spacecraft.* Patent No GB2334761B.
2. Shawyer R. The Development of a Microwave Engine for Spacecraft Propulsion. *JBIS* Vol.58 Suppl.1 (2005) 26–31.

6 "And Yet It Moves"

The static thrust measurements were initially carried out using a composite balance technique with the engine suspended from a crane. The majority of the 45 kg weight of the engine was supported on a spring balance, with residual weight supported on the 16 kg electronic balance via an interface plate with a preload screw adjustment. A diagram of the test rig is given in Figure 6.1. The preload could be set to give an overall resolution of 0.3 g. Calibration was carried out using standard weights. The calibration characteristic is given in Figure 6.2. The calibration factor is the result of the two spring constants forming a composite balance. The recorded balance data is divided by the calibration factor to give the calibrated data.

The stability of the calibration was, however, compromised by changes in ambient temperature, causing minor dimensional changes in the crane and suspension. These were minimised by applying thermal lagging to the critical areas. Stability was tested using a hot air blower with lagging applied until sufficient improvement was obtained to enable the test runs to be carried out successfully. Minor slope corrections were applied during data processing. Thermal calibration runs were also carried out in all three engine attitudes using the preheat function. This enabled the residual centre of gravity effects to be measured and compensated for in the processing of the results. Although the majority of the test runs were carried out with the engine thrust vector vertically up (i.e., the measured weight decreased while the engine was running), a number of runs were also carried out with the thrust vector vertically down. In these tests, the measured weight increased while the engine was running. Further tests were also carried out with the thrust vector horizontal when

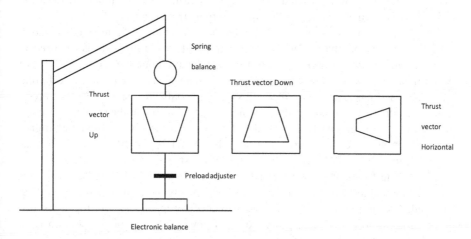

FIGURE 6.1 Diagram of the vertical test rig.

DOI: 10.1201/9781003456759-6

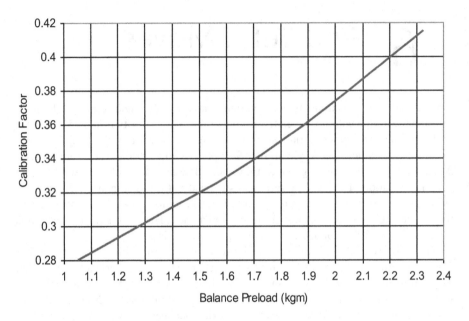

FIGURE 6.2 Vertical calibration factor.

the weight change recorded was due only to thermal effects. The vertical thrust test rig is illustrated in Figure 6.3 with the thrust vector horizontally forward.

To confirm more accurately the optimum resonance tuner setting, a number of swept resonance tuner runs were carried out. The stepper motor-driven, resonance tuner incorporates a feedback potentiometer, enabling accurate position telemetry to be recorded. The results for up, down and horizontal runs are given in Figure 6.4. The early test runs were carried out with non-optimised input tuner settings and showed considerable variation in the thrust outputs from run to run.

The swept tuner tests show that although the bandwidth of the magnetron output is 5 MHz (corresponding to 5 mm), careful tuning to within 0.1 mm is required to select the maximum component within the available output spectrum. A series of thrust measurement test runs were then carried out using a sequence of different input tuner settings. The input tuner stubs were manually set using multi-turn precision indicators on each stub. In this manner, the input match at the high-power magnetron setting was gradually improved to give increased levels of thrust with higher input powers to the engine. From an initial thrust of 1.28 g at 193 watts, thrust was increased to 8.33 g at 744 watts. The sequence of tests is given in Table 6.1.

The raw test data from test run DEV 71 is illustrated in Figure 6.5. This run recorded the highest specific thrust of 178 mN/kW, giving a calculated Q of 31,569. The high Q results from the low loading effect of the input tuner setting. This also gives a high ratio of reflected power to input power. The magnetron was on for a period of 30 seconds in this test run.

The effect of the magnetron warm-up frequency shift can be seen in this run. The thrust builds up throughout the 30 seconds as the frequency of the major component of the magnetron output is pulled into tune with the resonator. Separate frequency

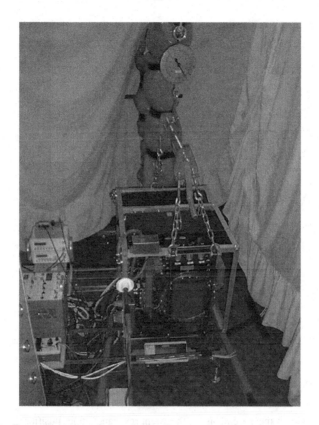

FIGURE 6.3 Demonstrator engine static test rig.

measurement test runs have shown different warm-up shifts for different magnetrons and for different input tuner settings. All thrust measurement runs have shown the effect of some degree of warm-up frequency shift.

The thrust data given in Table 6.1 is the mean of the processed telemetry data. The processing removes any test-rig response slope. A typical slope can be seen in the thrust plot in Figure 6.5. The data is then divided by the calibration factor, which varies with the preload on the balance and is given in Figure 6.2. Finally, the mean preheat response is subtracted.

A total of 95 vertical thrust test runs were carried out, most of them with the engine running for periods of 60 seconds. Many of the test runs were carried out without the engine suspended to enable adjustment of the input tuner stubs whilst monitoring input and reflected power. Detailed graphs showing the results of a number of the test runs are given in the technical report submitted to the UK government [1], which can be accessed on the emdrive.com website.

Following the vertical thrust tests, the test rig was modified to enable thrust to be measured horizontally. In this configuration, the engine was suspended from the crane, and a 90° thrust transfer rig was used to transfer the horizontal thrust into a vertical force, which was measured directly using the 110 g electronic balance. Calibration was carried out by attaching a cord at the centreline of thrust and running

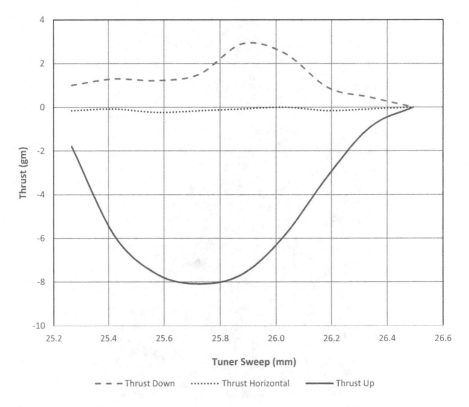

FIGURE 6.4 Thrust measurements for the swept resonance tuner position.

it over a pulley to a weight pan. The rig was then calibrated using standard balance weights. A diagram of the test rig is given in Figure 6.6.

Calibration was complicated by the fact that the centre of gravity of the engine was not on the centre line of thrust. This caused a rotational torque to be set up, and the actual thrust measured was a component of the thrust produced. To prevent the balance reaction from causing centre of gravity movement, a pivot bar was mounted under the centre of gravity of the engine. The value of the measured component of thrust was calculated from the geometry of the rig, as shown in Figure 6.7.

The thrust acting on the transfer jig at point P results in torque around point C. It can be resolved into two orthogonal components, a and b. The force measured by the electronic balance via the transfer jig is the component t. The resulting calibration factor of 0.074, calculated from the as-built engine dimensions, was checked against the thrust for a similar test run with the vertical test rig. (i.e., comparison of DEV 91 and DEV 107 gives 4.62 g for 534 W input power, compared to 4.29 g for 498 W). This gives exactly the same specific thrust for both tests. The direct use of a high-resolution balance also gave a fast measurement response, enabling a good correlation to be seen between thrust and input power characteristics. The test rig is illustrated in Figure 6.8.

TABLE 6.1
Thrust Measurements for Varying Input Tuner Settings

Test ref	Thrust vector	Input tuner settings			Output power (Watts)	Mean thrust (gm)	Specific thrust (mN/kW)	Q
		S1	S2	S3				
DEV 60	Up	874	394	0	193	1.28	65	11,563
DEV 62	Up	630	0	0	404	3.73	91	16,097
DEV 65	Up	524	28	472	386	1.79	45	8,085
DEV 71	Up	630	1400	802	338	6.12	178	31,569
DEV 74	Up	918	1400	584	552	4.18	74	13,203
DEV 77	Up	630	1400	584	561	4.05	71	12,587
DEV 78	Up	778	1400	584	543	4.34	78	13,935
DEV 81	Up	848	1400	584	572	3.4	58	10,363
DEV 82	Up	778	1400	693	443	3.8	84	14,956
DEV 83	Up	778	1400	471	697	3.94	55	9,856
DEV 84	Up	778	1400	358	914	4.12	44	7,859
DEV 85	Up	630	1400	358	887	4.31	48	8,472
DEV 88	Down	630	1400	358	820	5.54	66	11,779
DEV 89	Down	630	1400	358	744	8.33	110	19,521
DEV 91	Down	778	1400	584	534	4.62	85	15,084
DEV 92	Down	778	1400	584	512	5.05	97	17,197
DEV 94	Horizontal	778	1400	584	530	0.19		
DEV 95	Horizontal	778	1400	584	516	0.37		

TABLE 6.2
Horizontal Test Runs

Test ref DEV	Input tuner settings			Resonance tuner mm	Input power (Watts)	Mean thrust (gm)	Spec thrust (mN/kW)	Q
	S1	S2	S3					
107	778	1400	584	15.5	498	4.29	85	15,019
120	778	1400	584	15.2	498	4.87	96	17,050
126	780	1400	70	15.2	357	4.87	134	23,784
130	468	1400	70	15.2	302	2.85	93	16,454
132	468	0	70	15.2	151	3.29	214	37,988

The results of the horizontal tests are summarised in Table 6.2. The object of the sequence was to further investigate the envelope of impedance match between tuner and resonator and between tuner and magnetron. With the magnetron power supply set to constant power settings, the results of the test runs varied from a maximum thrust of 4.87 g for 498 W input power to 3.29 g for 151 W input power. These values correspond to Q values of 17,050 and 37,988, respectively. They illustrate the increase in specific thrust that can be achieved with a low input loading on the resonator.

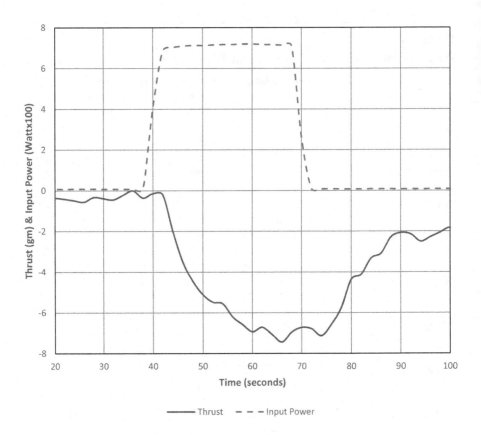

FIGURE 6.5   Test run DEV 71. Thrust direction UP.

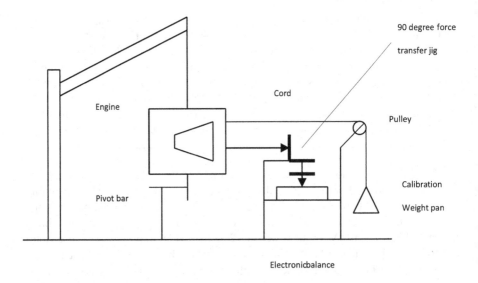

FIGURE 6.6   Diagram of the horizontal thrust measurement rig.

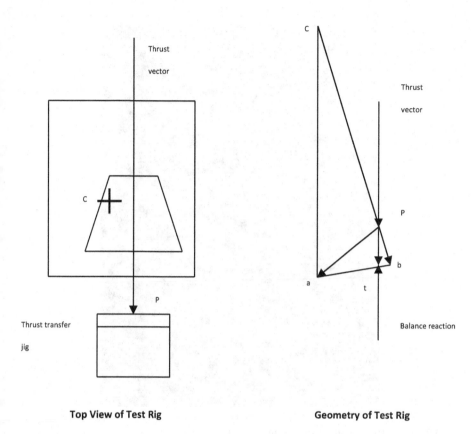

Top View of Test Rig   Geometry of Test Rig

FIGURE 6.7   Geometry of a horizontal test rig.

Note also the increase in Q, from 15,019 in DEV 107 to 17,050 in DEV 120, when the resonant tuner depth is decreased by 0.3 mm. This illustrates the need to track any frequency drift of the main spectral component of the magnetron output. Over the range of tests quoted, a drift of approximately 4 MHz, from 2,451 MHz to 2,447 MHz, was measured. This required a resonance tuner depth increase from 11.4 mm to 15.2 mm to maintain the optimum resonance setting. Figure 6.9 shows the results for DEV 132, illustrating the highest Q achieved (37,988). A final test, DEV 134, was carried out, and DEV 132 was repeated with the balance locked. This test confirmed that EMC effects on the balance data are negligible. Once again, a set of detailed graphs showing the results of a number of the horizontal test runs are given in the technical report [1].

At this point, we were running out of schedule and funds but had not yet demonstrated the engine on the rotary test rig. The construction of the test rig had been considerably delayed due to a slip in the delivery of the air bearing. We therefore produced a technical report on the work to date [1] and, as with the feasibility study, had it reviewed by John Spiller [2]. The report was delivered in July 2006 and accepted by DTI, eventually enabling the final DTI payment to be received. Fortunately, by this time, our second major sponsor had arrived on the scene. Paul Young, an engineer and entrepreneur who had a strong belief in the future of the UK space industry,

FIGURE 6.8   Horizontal thrust measurement rig.

became a major shareholder in the company in return for a significant investment. This funding enabled the critical acceleration tests to be carried out. Paul's continuing financial support was to keep us working for the next five years and to enable the completion of the demonstrator programme. We were also able to carry out a superconducting experiment and start the Flight Thruster programme.

After a full peer review of the technical report by the DTI and their independent experts, the following conclusions were reached:

Conclusions from the Demonstrator Engine Technical Report to DTI.

1. The microwave design was based on updated software used in the design of the experimental thrusters.
2. The microwave design was validated by a series of small signal sweep tests.
3. The thermal compensation design was validated by thermal testing of the engine under small signal conditions.
4. The thermal subsystem design was initially tested on a thermal test rig and was then shown to provide sufficient cooling to enable high-power tests to be carried out.
5. High-power sweep tests showed good agreement with the small-signal sweep tests.

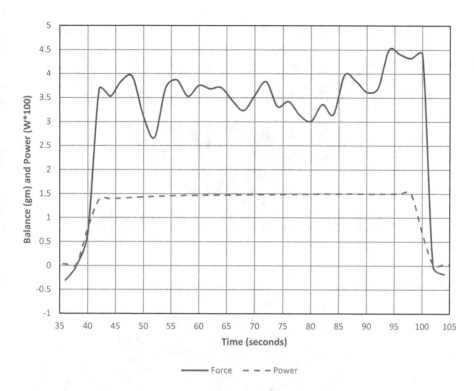

FIGURE 6.9 Test run DEV 132.

6. Vertical test rig development resulted in a fully calibrated rig capable of measuring engine thrust to a resolution of 0.3 g.

7. Vertical sweep tests confirmed that thrust peaked at the design resonance point.

8. Thrust peaks were measured in the correct direction with the engine in both up and down configurations. No vertical thrust was measured with the engine horizontally.

9. A series of vertical thrust tests with a fixed resonance setting but with different input tuner settings gave an envelope of performance covered by:
   Min. 1.28 g thrust for 193 watts of input power.
   Max. 8.33 g thrust for 744 watts of input power.

10. The vertical thrust tests included up, down and horizontal engine configurations. In each case, the thrust was measured in the correct direction, with no vertical thrust being measured with the engine horizontally.

11. A horizontal test rig was developed, which enabled calibrated thrust measurements to be made with the engine in a horizontal position.

12. The input tuning tests were continued on the horizontal test rig and gave a maximum specific thrust result of 214 mN/kW. The measured results were 3.29 g thrust for 151 watts of input power.

13. It is concluded that the demonstrator engine has provided further evidence to validate the theory and has enabled significant progress to be made towards a flight engine design.

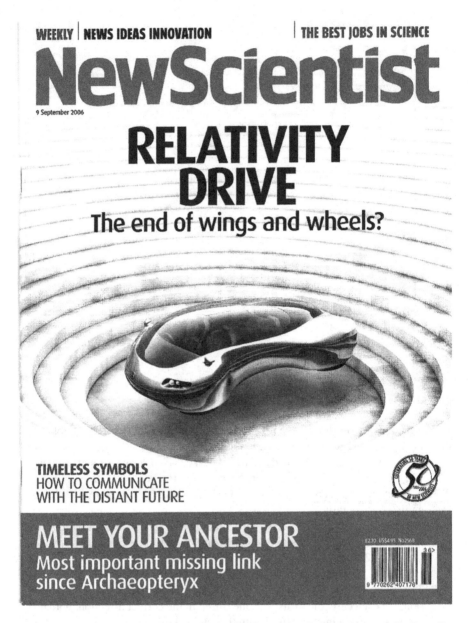

FIGURE 6.10   Front cover of the new scientist, September 9, 2006. (Artwork by Benedict Campbell)

One month prior to the submission of this technical report, a cover article with the title *Relativity Drive – The end of wings and wheels?* was published by New Scientist [3]. The cover showed an impressive artist's impression of a flying car, as shown in Figure 6.10.

This serious attempt to explain EmDrive and its implications caused some controversy. Although we received an overwhelmingly positive response, a small but very vocal minority objected to the concept. Whilst none of the critics bothered to understand the physics or to do the math there was the accusation that the whole idea was impossible and that the project had been a complete waste of taxpayers' money. A New Scientist blog was started, and controversy raged between an internet troll, supported by the ex-Marconi director, and a number of more technically literate contributors. We looked on with incredulity and amusement. The DTI officials contacted the appropriate people to be informed that the troll was well known to them for his previous diatribes on a number of topics, including the assassination of John F. Kennedy. It all eventually led to a written question in the House of Commons, which was robustly answered by the then Minister for Trade and Industry, Margaret Hodge. Hansard, the official record of UK parliamentary debates, gives the following information:

ELECTROMAGNETIC RELATIVITY DRIVE

Alan Duncan: To ask the Secretary of State for Trade and Industry how much his Department has provided for the electromagnetic relativity drive design proposed by Roger Shawyer; and from what budget funding has been drawn. [103254]
Margaret Hodge [holding answer November 27, 2006]: Awards have been made to Satellite Propulsion Research Ltd. from the DTI's Small Firms and Enterprise budget.
   July 2001—£43,809 paid.

   A feasibility study into the application of innovative microwave thruster technology for satellite propulsion. The study involved development of an experimental thruster followed by independent tests and evaluation.

August 2003—£81,291 total grant awarded, £68,399 paid to date.

   A follow-on from the above project, to design and develop a demonstration model engine. To be tested on a dynamic test rig, to demonstrate continuous thrust and the conversion of thrust into kinetic energy.

Both grants were awarded against the criteria of the DTI's Smart scheme, which was designed to help fund pioneering and risky R&D projects in small and medium enterprises. Highly qualified technical experts and academics carried out an assessment on behalf of the department.

This put an end to the debate but had the unexpected effect of alerting both the US defence agencies and a prominent Chinese professor to the concept. Both of these parties had important parts to play in the subsequent story. However, it was clearly necessary to finish the project by demonstrating the acceleration of the engine. The total weight of the demonstrator engine, power supplies and instrumentation was approximately 100 kg. To obtain consistent acceleration measurements over 360 degrees of rotation, the friction torque of the bearing would ideally result in a force that was less than one-tenth of the thrust produced by the engine. With a maximum thrust of 8 g, this was a friction force of less than 0.8 g. The requirement for a bearing friction force that was 125,000 times less than the load it was carrying was a major challenge. Further development work on the dynamic test rig included a new air compressor and improved dynamic balancing. Also, a programme of work was carried out to develop a new input feed assembly. The object was to improve the

FIGURE 6.11   Dynamic test rig.

magnetron–resonator impedance match and increase the input power to the engine whilst maintaining the optimum Q loading. This would increase the output thrust. The work was made easier following the procurement of new microwave instrumentation, including a precision sweep generator. The dynamic test rig is illustrated in Figure 6.11.

The test rig comprised a carefully levelled and balanced beam, supporting the engine on one side and an instrument platform on the other. This beam was mounted on a Loadpoint 10-inch air-bearing type D03099, mounted in a subframe. Mains power was supplied to the centre of the beam via flexible wires coiled such that the operation of the test rig was limited to a small number of turns. The total beam weight was 100 kg. The tests therefore simulated the engine moving a 100-kg spacecraft in weightless conditions. The programme included acceleration and deceleration runs in both directions. A major problem in carrying out these test runs was the variation in friction torque in the air bearing due to the variation in humidity and temperature of the ambient air throughout a typical day. Calibration runs to determine the friction torque were therefore carried out prior to each dynamic test run. These used a cord and pulley together with standard weights to provide the calibration force along the thrust axis. This proved successful during the horizontal thrust measurements and was used to provide a preload for some dynamic tests, where the friction torque proved too high to obtain movement without the necessary additional force. All test and calibration runs were recorded on video, enabling performance results to be processed from an analysis of the angular movement and elapsed time data. For each engine test, telemetry data was recorded on the laptop attached to the instrument platform. Telemetry data included power, current and temperature measurements.

A summary of the test runs is given in Table 6.3.

The video data processing was a time-consuming activity, and therefore the number of runs was limited, but tests were carried out from a static start, from a moving start, and in both forward and reverse directions. The conclusions were, however, quite clear:

TABLE 6.3
Summary of Dynamic Test

DEM No	Engine direction	Start	Preload (gm)	friction (gm)	Thrust (gm)	Power (W)	Time (s)	Sp thrust (mN/kW)
185	Forward	Static	10.64	15.96	7	423	126	162
186	Forward	Static	3.99	14.41	11.87	293	97	397
187	Forward	Moving	0	9.21	9.99	299	152	328
188	Forward	Static	0	9.21	9.77	334	88	287
190	Forward	Static	0	7.9	7.32	343	75	209
191	Reverse	Moving	0	4.17	6.51	315	166	203
192	Reverse	Moving	0	4.17	6.28	344	149	179
194	Reverse	Static	3.12	5.63	3.21	260	169	121
196	Reverse	Moving	3.12	2.95	8.72	251	107	341

Yes, it moves, yes, it is in the correct direction according to Newton's third law, and yes, the calculated thrust from the acceleration, using Newton's second law, is in reasonable agreement with the measured static thrusts.

The mean specific thrust from all the dynamic test runs was 248 mN/kW, compared to the maximum specific thrust of 214 mN/kW from the horizontal static thrust measurements. A comparison of the specific thrusts recorded in Tables 6.1–6.3 shows the steady improvement in thrust as experience is gained in the tuning necessary to achieve an improved input impedance match.

It was agreed that the raw video of DEM 188 would be released to a limited number of interested parties, including the US defence agencies. The feedback that we later received suggested that heated discussions had taken place between USAF, the National Security Agency (NSA) and the Defence Advanced Research Projects Agency (DARPA) personnel and that they resembled *guys in trenches throwing grenades at each other*. Clearly, the controversy around EmDrive had risen way above the New Scientist blog participants. The old saying that there is no such thing as bad publicity obviously applies here. The video was finally released into the public domain in June 2015 and is available for download from the SPR Ltd. website [4]. The DEM 188 video illustrates a number of interesting points, the first being that the high Q of the resonator gives rise to very high field strengths within the thruster. They equate to an instantaneous power level of 17 MW. Signal leakage causes EMC effects within the fixed video camera. This leads to the apparent vertical movements once resonance starts within the thruster. Also, the audio track, which has a background noise of the compressor providing air to the bearing, includes comments indicating "on" and "off" of the magnetron and the intermediate frequency (i.f.) signal frequency. The i.f. frequency is the intermediate frequency used by a down-converter monitoring the signal leakage and read directly from the spectrum analyser. The resonant frequency of the engine gives an equivalent i.f. frequency of 23 MHz. The magnetron frequency, and thus the leakage frequency, changes as the magnetron warms up and only stabilises once the magnetron has locked to the resonant frequency of the cavity. The engine only starts to accelerate when the magnetron frequency locks to the resonant frequency. Importantly, this illustrates that it is not electromagnetic or thermal spurious forces causing acceleration. The change in i.f. frequency is plotted for DEM 188 in Figure 6.12, and the input power and engine velocity in Figure 6.13.

Comparing Figures 6.12 and 6.13, it is clear that the engine only starts to move when the frequency of the magnetron starts to pull in towards the resonant frequency of the cavity around 125 seconds. Acceleration is steep, with the rate variable with power input, until around 210 seconds, when the power is turned off. The acceleration then decreases but is at a fixed rate until around 255 seconds, when the engine rotation slows due to friction torque. This second period of acceleration when the power was turned off remained a mystery until an experienced rocket engineer said it looked just like the effect of propellant movement following a stage separation during a typical launch. Propellant is forced to the bottom of the tank during initial acceleration, but once the first stage rocket stops, propellant is free to move up the tank due to its own inertia and provides a small acceleration to the vehicle during the drift period until the next stage ignites. Indeed, in some cases, it is necessary to fire small ullage motors to settle the propellants in the next stage. The demonstrator engine contains almost 5 kg of water coolant, and it was suggested that this was what was giving the mysterious period of acceleration. A few calibration runs seemed to confirm the effect.

The demonstrator engine had done its job but was subject to the understandable criticism that it was never going to fly in this configuration. Significant improvements in mass reduction and repeatability of results were going to be required in the next stage of development. But this comment, attributed to Galileo, also applied to EmDrive. In spite of all the controversy, *"and yet it moves"*.

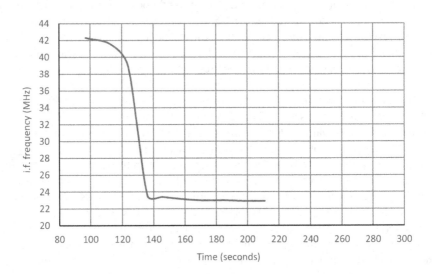

FIGURE 6.12   IF Frequency for test run DEM 188.

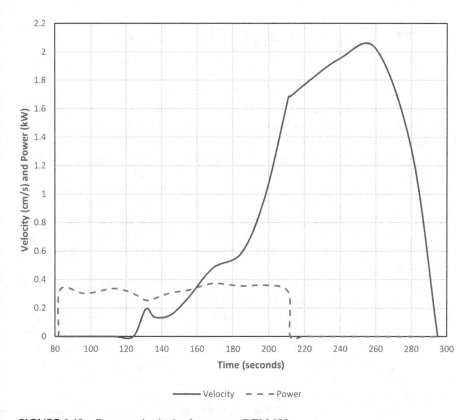

FIGURE 6.13   Power and velocity for test run DEM 188.

REFERENCES

1. Technical Report on the Development of a Microwave Engine for Satellite Propulsion. Issue 2. July 2006. R&D Development Project Ref.1939. SEEDA. DTI.
2. Spiller J W. Review of Technical Report on the Development of a Microwave Engine for Satellite Propulsion.JWS-SPR-TN-018. 27 August 2006.
3. Mullins J. Fly by light. Cover story New Scientist 9 September 2006.
4. Emdrive - Home. June 2015. Dynamic Test. Full DM test 188. www.emdrive.com.

7 The Flight Thruster

The flight thruster concept arose during a number of discussions carried out with different American aerospace companies during our US visit in September 2006. The meetings were first set up by our newest shareholder, Philip Owen, who had become our Business Development Manager. Philip, an ex-IBM executive, had a wide range of contacts in government and industry in both the UK and America. One of his operating bases was the Reform Club in London, so the introductions and discussions in such elegant surroundings brought a new level of sophistication to our business dealings. The discussions resulted in significant interest being shown by Boeing, and following a number of teleconferences, an export licence was applied for, to enable EmDrive technology to be transferred to America. The licence was granted by the UK in January 2008.

It was clear that a flight representative thruster would need to be the next major step in our development programme, and much thought went into its outline specification. Typical UK space equipment programmes follow the traditional model philosophy of the Breadboard, engineering model, qualification model and finally the flight model. The American philosophy was more direct, with one US organisation declaring that they would be prepared to fly an engineering model, if suitable. A major decision had to be made about what microwave power source would be used. Obtaining a magnetron suitable for flight use did not seem possible, so the choice was between a travelling wave tube amplifier (TWTA) and a solid-state power amplifier (SSPA). To keep the thruster cavity to reasonable dimensions while retaining the TEO1 operating mode, which had been successful in the demonstrator engine, it would be necessary to increase the operating frequency from 2,450 MHz. However, care had to be taken over what frequency would be acceptable for electromagnetic compatibility (EMC) purposes on a typical test satellite. In addition, it was felt that if we could design the thruster around existing flight-qualified microwave amplifiers, the design stood a better chance of being flown. At that time, an SSPA of sufficiently high frequency and power output represented a high-mass, high-cost route and would not provide attractive efficiencies. The choice thus rapidly narrowed down to multiple C-Band TWTAs with up to 150 W output, which were available as flight-qualified units. After some frequency analysis, it was felt that an operating frequency of nominally 3,850 MHz would be acceptable for integration into a typical communications satellite. Mechanical integration issues were addressed, resulting in a requirement for a co-axial cable input rather than the waveguide input of the demonstrator engine. The thermal interface with the satellite was defined as a maximum thermal flux density of 15,000 W/m^2 averaged over the baseplate. A qualification temperature range at the baseplate was specified as $-15°$C to $+85°$C, with a cold start of $-30°$C. The frequency and thermal interface requirements led to an outline cavity design with a maximum baseplate diameter of 265 mm and a maximum power dissipation of 600 W. To meet reliability requirements, at least two TWTAs were required, and as

DOI: 10.1201/9781003456759-7

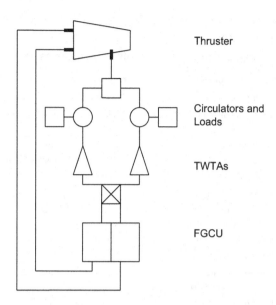

Thruster

Circulators and
Loads

TWTAs

FGCU

FIGURE 7.1   Functional block diagram of the flight engine.

the identified flight TWTA maximum output was 150 W, the operating output performance was specified at a microwave input power of 300 W. The output thrust at this power level was specified as 85 mN. The outline specification was issued as a formal document and underwent a number of modifications as discussions continued with both government and industry representatives.

Clearly, to achieve a stable output thrust over the full operating temperature range, accurate control of the input frequency would be required. It was proposed that a dual-redundancy frequency generator and control unit be adopted, with the amplitude of a monitor signal being used as feedback in a frequency control loop. The frequency generator and control unit output signal were to be capable of varying from zero to a maximum level, such that the total output from the two TWTAs was 300 W. A functional block diagram of the full flight engine is given in Figure 7.1.

The basic cavity design was carried out using the same Excel-based design software that was used on the experimental and demonstrator cavities. By this stage, the model was using axial length increments of 0.1 mm and had been fully validated against measured data. This enabled the cavity geometry and dimensions to be calculated with great confidence. The impedance and position of the input and detector circuits could also be determined, and the theoretical, unloaded Q could be calculated. By running the model with dimensional changes over the specified temperature range, the predicted frequency shift could be predicted. The model was also run for different modes to establish the frequency margins between the desired TE013 mode and the nearest unwanted modes.

The design is summarised in the output file given in Table 7.1, together with the E and H field plots given in Figures 7.2 and 7.3 and the impedance plot in Figure 7.4. The E and H field plots were used to establish the positions of the input loop (maximum H field), and detector probe (maximum E field) and vent hole (minimum E field).

TABLE 7.1
Design summary for the flight thruster cavity

Cavity Geometry	Circular	
Environment	Vacuum	
Major dimension D1	200	mm
Minor dimension D2	97	mm
Taper length L1	161.5	mm
Cone angle	17.69	degrees
Propagation mode	TE01	
No. of half wavelengths p	3	
Resonant frequency	3873.6	MHz
Major guide wavelength	87.75	mm
Minor guide wavelength	333.61	mm
Design factor	0.8168	
Cut-off diameter	94.36	mm
Cut-off frequency @ D2	3767	MHz
Input position	45	mm
Input impedance	314.4	Ohms
Detector position	121.8	mm
Detector impedance	239.6	Ohms
Cavity material	Silver-plated aluminium	
Theoretical unloaded Q	73,243	
Theoretical static thrust	399	mN/kW
Unloaded 3 dB bandwidth	53	kHz

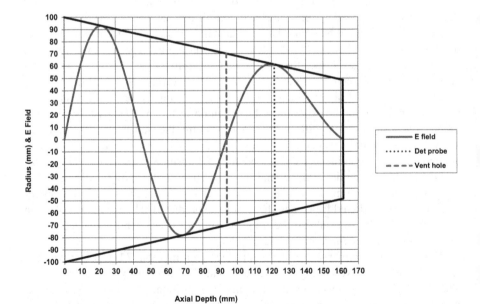

FIGURE 7.2 E-field phase response for the flight cavity.

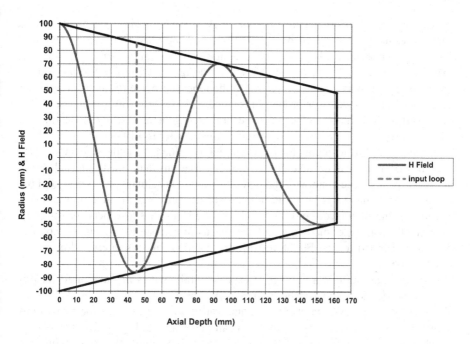

FIGURE 7.3   H-field phase response for the flight cavity.

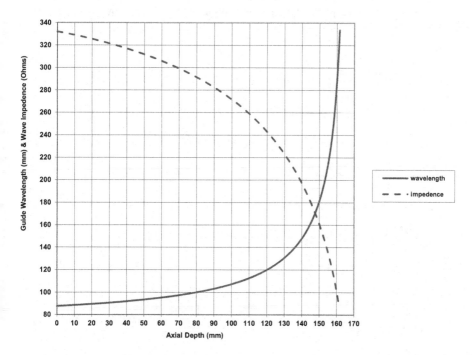

FIGURE 7.4   Guide wavelength and wave impedance for the flight cavity.

The thruster cavity is machined from an AlMgSi 6061 alloy and is assembled from three main components: baseplate, body and top plate. These are fixed by a total of 72 socket-head M4 screws. The flange thickness and flatness ensure good EMC performance. The input circuit is a loop design based on an N-type connector. A tuning screw is mounted adjacent to the loop to provide resonance tuning for the input circuit. A detector probe is provided to monitor cavity resonance and is assembled from an SMA connector. A general arrangement drawing of the thruster is given in Figure 7.5. The measured mass of the thruster is 3.193 kg.

A major factor in achieving the required high Q value for the thruster is the wavefront distortion that can be generated by minor variations in the basic internal geometry. Clearly, a simple cone with flat end plates will cause major distortion, and the actual internal shape has been developed over the previous thruster programmes. The machining tolerance required to achieve the shape has also been investigated theoretically and experimentally, and the results are illustrated in Figure 7.6. The tolerance specified for the Flight Test Model (FTM) thruster's internal dimensions was +/–0.05 mm.

It is worth noting that a number of experimental cavities manufactured for private and academic research programmes either did not have machining tolerances specified, or when the researchers were questioned, they replied that the dimensions were within *"about a couple of mm"*! As can be seen from Figure 7.6, the chances of achieving a cavity Q above 5,000 were vanishingly small. But they claimed to have measured Q values above 30,000 and input return losses of 45 dB using their highly expensive vector network analysers. In reality, of course, they were measuring the input circuit response, which was completely unmatched by the cavity wave

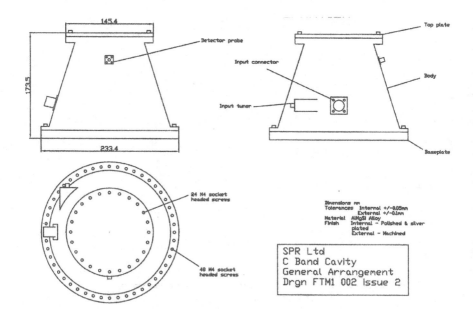

FIGURE 7.5 Flight thruster drawing.

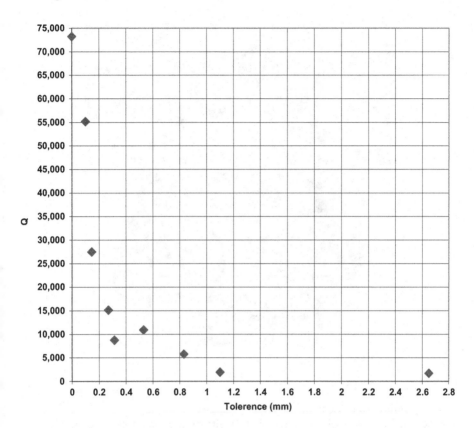

FIGURE 7.6   Dimensional tolerance effect.

impedance. Figure 7.4 apparently meant nothing to them. Whilst not totally agreeing with the comment from a colleague *"that they may just have well have tried to get an empty bean can to resonate"*, it remained desperately hard for me to persuade them to carry out the correct design, manufacturing and tuning procedures. One of the main steps in the correct set of procedures is the manufacture of a breadboard before the considerable expense of precision machining and highly accurate dimensional measurement of the cavity is undertaken. In the case of the flight thruster, this took the form of a flight electrical model (FEM) cavity.

The FEM was manufactured by fabricating the body and end plates from copper sheet, with steel supporting plates used to provide mechanical stability. The input assembly was based on a standard N-type co-axial connector, with an SMA connector used for the detector probe. The FEM thruster is illustrated in Figure 7.7.

A simple beam balance force measurement rig was built, enabling a programme of development tests to be carried out with the thruster driven by a 40-watt SSPA.

FEM development tests included:

1. Confirmation of the resonant frequency
2. Measurement of loaded Q
3. Development of the input circuit and tuning assembly

FIGURE 7.7   Flight electrical model.

4. Optimisation of input and detector positions
5. Characterisation of the variation of Q with dimensional tolerance
6. Development of test bench design
7. Measurement of thrust at low input power

A total of 145 test runs were carried out with the FEM. The fabricated construction enabled connector positions to be easily varied and different tuning assemblies to be tested. A shimming technique was developed to enable the variation of Q with dimensional tolerance to be fully investigated. The thrust measurements confirmed that reliable, repeatable measurements at low input power could not be guaranteed with a co-axial cable linking the fixed source to the suspended thruster.

Whilst the early design work on the flight thruster was taking place, an invitation to visit China was received. This was from Professor Yang Juan of the North Western Polytechnical University (NWPU) at Xi'an, who had carried out her own theoretical research since being alerted to EmDrive by the controversy caused by the New Scientist article. Advice was taken from the appropriate UK government agency, and it was agreed that a visit to discuss theory and give a couple of lectures was appropriate. The main aim, however, was to obtain information as to the progress of China in actual EmDrive development. The visit was certainly an eye-opener to the quality and size of NWPU. Having become familiar with a number of US and UK government research establishments and universities, I judged NWPU to be similar to The Aerospace Corporation in El Segundo, only much larger. It certainly dwarfed any European university I had visited. Professor Yang's group worked in an electrical propulsion laboratory, which made the equivalent facilities at Farnborough look antiquated. They routinely tested flight-quality microwave ion thrusters in thermal vacuum chambers, using power levels up to 2.5 kW, and did work for the Japanese Space Agency on the microwave thrusters for their Hayabusa asteroid mission. One thing that immediately became clear was that they understood the engineering challenges of high-Q, high-power microwave cavities. As Professor Yang said, their cavity design was not unlike my EmDrive cavity, except that it had a hole in it to allow the plasma to escape. They also understood the limitations of finite element analysis, and one of the post-grads had actually disassembled a commercial software package and reassembled it with the equations necessary to correct for guide wavelength. He claimed that it would indeed illustrate a collapse of field structure when cut-off conditions were met, unlike the original western software. There was no doubt that once they started experimental work, progress was likely to be rapid. My observations were duly reported back to the UK government.

A few months later, I was introduced to Colonel Mike Smith of the USAF, known by his call sign Coyote. He had been responsible for a Solar Power Satellite Study while working for the National Security Space Office (NSSO) at the National Security Agency (NSA) in Washington. He immediately understood the implications of the superconducting EmDrive thruster, which we had been working on in parallel with the flight thruster programme. Coyote then began the formalities necessary for a meeting at the Pentagon. He actually described an EmDrive-powered launch vehicle as *"a space elevator without the cables"*. I was suitably impressed. Our work up to that point had been reported at the IAC08 conference in Glasgow [1]. In the exhibition associated with the conference, we displayed a 2 m aerodynamic model of an EmDrive-powered space plane that had undergone flight tests in Gibraltar, although only powered by model aircraft turbines. We actually woke up in the exhibition hall by running up the engines for a brief period, causing much interest. The spaceplane model is illustrated in Figure 7.8.

As we already had an export licence in place for the US, agreement for a meeting at the Pentagon was given by the UK authorities. We arrived at the Pentagon on December 10, 2008, to be met by a US Marine officer in full uniform and a stern-looking security detail on the front desk. On a previous visit, as a MoD contractor, I had sailed through the identification formalities. But not this time. The identification requirements were two documents with photos, which usually meant a passport and a driving licence. This was a problem, as my licence was an old-fashioned British piece of green paper, now very dog-eared. In desperation, I produced my pensioner's bus pass, which was passed amongst the desk clerks to much amusement. The meeting itself was no less intimidating, with representatives of the USAF, the US Marine Corps (USMC), The Royal Australian Air Force (RAAF), DARPA, the National Aeronautics and Space Administration (NASA) and the National Security Space Office (NSSO) submitting us to heavy questioning under the chairmanship of the Director of NSSO. The next day we had a further meeting with DARPA, chaired by the Assistant Director, where we discovered that Boeing was already working on their own EmDrive programme, sponsored by DARPA.

FIGURE 7.8   EmDrive space-plane aerodynamic model.

A number of actions were taken to obtain UK government support to ensure that we were not left out of the ongoing work. This eventually resulted in a Technology Assistance Agreement being set up between Boeing and SPR Ltd., with the agreement of the US State Department and UK MoD. A contract with Boeing for their Phantom works at Huntington Beach was started in November 2009. The work included reviewing Boeing's initial EmDrive design, preparing a detailed design against Boeing's requirement specification, and reviewing the test data for this design. Unfortunately, security implications prevented Boeing from sending us their results. To complete the contract, we agreed to provide our test data for the flight thruster, which was not too dissimilar to the new Boeing design. In an effort to continue the US/UK collaboration, a licence agreement was negotiated, but before this was signed off, we were told that Boeing was no longer prepared to licence the technology but simply wanted to buy SPR Ltd. and all the EmDrive intellectual property rights. This rather startling change of tack was not acceptable, and so we continued with our own development programme.

Initial development tests on the flight thruster, illustrated in Figure 7.9, covered the effect of high-power transfer in the co-axial cable link on spurious forces. This early work resulted in a redesign of the force measurement rig to incorporate a free waveguide coupling between source and thruster. This mechanically decouples the fixed and suspended parts of the force rig. The coupling provided a force resolution approaching the 0.1 g resolution of the electronic balance. However, considerable design and development work was required to achieve this resolution while minimising transmission loss and leakage levels. A block diagram of the test bench is given in Figure 7.10.

The digitally synthesised signal generator gave a calibrated output signal that could be stepped in amplitude and frequency via front panel controls. TWTA output power was monitored by a digital power metre, which was connected to the main computer via a USB port. Detected power in the cavity and reflected power at the

FIGURE 7.9 The flight thruster.

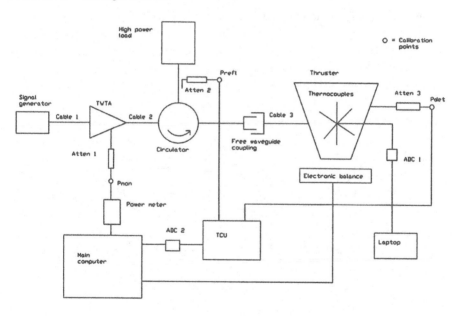

FIGURE 7.10 Block diagram of the test bench.

cross coupler were monitored via attenuators at ports on the Telemetry and Control Unit (TCU), originally built for the demonstrator engine. The TCU provided isolated and calibrated data input to the main computer via an analogue to digital converter (ADC), as well as a digital display of both power levels. Force data was monitored at the electronic balance via a serial link to the main computer. The electronic balance was electrically isolated by being powered by a rechargeable battery power supply. For EMC purposes, the thruster temperatures were monitored on a separate laptop computer via an ADC unit. Data was transferred to the main computer via disk after

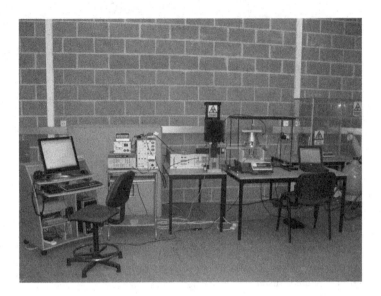

FIGURE 7.11   General view of the test bench.

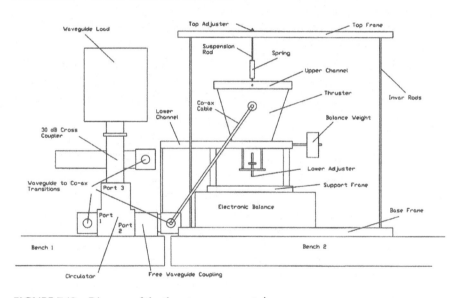

FIGURE 7.12   Diagram of the thrust measurement rig.

each test run. All data from a test run was recorded on an Excel spreadsheet running
on the main computer. Once calibration data had been entered, the spreadsheet gave
immediate graphical output for test run analysis. A general view of the test bench is
given in Figure 7.11.

The thrust measurement rig is a composite balance with the thruster suspended
between an extension spring with a spring constant of 2.8 kg/mm and an electronic
balance with a spring constant of 13.9 kg/mm. A diagram of the rig is given in
Figure 7.12, and it is illustrated in Figure 7.13.

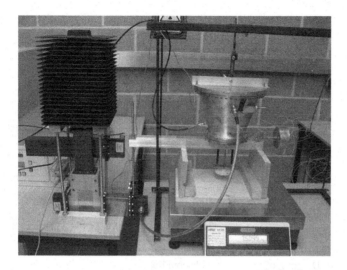

FIGURE 7.13 View of the thrust measurement rig.

The spring is hung from a steel top frame, which is supported by invar rods fitted to a steel base frame. The invar rods minimise spurious force measurements due to ambient temperature changes. The rig is mounted on a separate test bench from that supporting the TWTA to minimise force measurement noise from the cooling fans in the TWTA and signal generator. The thruster itself can be mounted in two configurations: thrust vector up or thrust vector down. It is mounted between two aluminium channels, the upper channel being connected to the suspension spring. One end of the lower channel supports a waveguide-to-co-axial transition, mounted on two steel rods. At the other end of the channel, a balance weight is mounted to provide adjustment of the horizontal attitude of the beam and thruster. Directly under the central axis of the thruster, a lower adjuster screw is attached to the lower channel.

Spurious forces are minimised by transmitting the microwave power from port 2 of the circulator via a free waveguide coupling, where there is no mechanical coupling between the fixed and floating waveguide elements. It was found, during the development tests, that with careful mounting of a small flexible co-axial cable, it was possible to monitor detected power during a thrust measurement test run without producing a measurable spurious force. Similarly, the thermocouple connections and a ground connection could be made to the thruster without incurring spurious forces. The weight of the thruster can be distributed between the suspension spring and the electronic balance by either using a screw adjuster on the top suspension rod, the lower adjuster, or both. During the development test programme it was found that the most stable method of adjustment was to dispense with the lower adjustment screw and use a wooden support frame. The pre-load (i.e., the weight supported by the electronic balance) was varied by adjusting the position of the top adjuster while monitoring the electronic balance readout. Due to the high spring constant of the electronic balance, the movement of the position of the "floating" side of the free waveguide coupling was minimal as the pre-load or measured force changed. The microwave power to the thruster input was transmitted via a co-axial cable, whose outer sheath was removed to

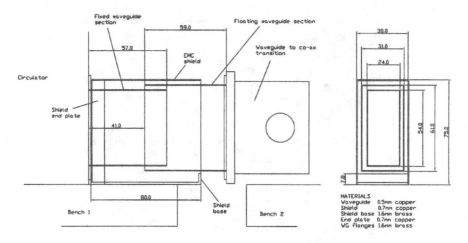

FIGURE 7.14   Diagram of free waveguide coupling.

improve cooling. Note that the "on" period of the high-power test runs was limited by the temperature rise in this co-axial cable. However, as the cable is mounted between two fixed points, temperature rise and thermal expansion had no measurable effect on the force measurements.

The free waveguide coupling is illustrated in Figure 7.14. The coupling consists of two waveguide sections fabricated from 0.5-mm sheet copper. These are soldered into brass flanges. The fixed section is mounted on the flange fittings of port 2 of the circulator. The floating section is mounted on the flange of the waveguide at the co-axial transition. The floating section is accurately aligned within the fixed section with a 3 mm gap on all sides. This gives a nominal quarter-wavelength choke section whose actual length is determined by the test. The length is optimised to give minimum transmission loss at the thruster's operating frequency. Leakage is minimised by surrounding the coupling with a two-part EMC shield comprising a brass shield base and a sheet copper shield. Note that it is important that both parts of the EMC shield are bonded to the electrical ground.

With high voltages present at the TWTA output, a comprehensive grounding system is used, which is checked visually before each test run. To ensure electrical noise on the telemetry data is kept to a minimum, shielded cables are used, with ground loops minimised by using a star layout. It was found necessary to completely isolate the temperature monitoring system to minimise noise, and an ambient thermocouple was used to monitor any level change during "on" periods. Calibration procedures included EMC checks at full power to eliminate any EMC errors in the measured data. The free waveguide coupling was a source of considerable signal leakage during high-power runs. The leakage field at the operator position was continuously monitored to ensure a safety level of $5\,mW/cm^2$ was not exceeded.

With a stable, high-resolution test bench, the flight thruster underwent a long series of tests to optimise the input circuit design and develop frequency tracking algorithms. Frequency tracking requirements were considerably more complex than originally foreseen due to the interaction of the thermal characteristics of the input

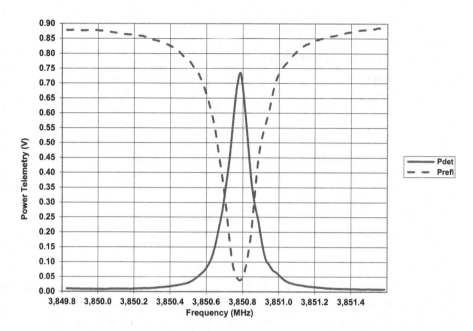

FIGURE 7.15   Flight thruster resonance plot.

circuit and the cavity itself. The change in input impedance as the input circuit rapidly warms up and the frequency effect of dimensional changes on the input loop itself creates an inherently unstable situation following the initial turn-on. As the cavity warms up, a second, more predictable, frequency-tracking phase dominates. A further, important objective of the thruster development programme was to increase the loaded Q of the cavity while maximising the return loss of the input circuit at resonance. This ensured that the operating point was close to the theoretical situation, where the loaded Q is half the unloaded Q, power transfer from the input circuit to the cavity is a maximum, and reflected power is a minimum. A total of 156 development test runs were carried out. At the end of this programme, 19 further performance tests were carried out, followed by two final calibration runs.

Tuning the cavity for maximum loaded Q was a lengthy, iterative process that occupied much of the development test programme. Quick Q measurements were taken by offsetting the input frequency from the resonant value until the detected power was halved and then calculating the 3 dB bandwidth. This then gave the loaded Q at the resonant frequency. Once the desired tuning point had been achieved, a plot of the detected power and reflected power against the input frequency was taken by stepping the frequency over the required range. For accurate Q determination, this was carried out for both increasing and decreasing frequency steps, with a mean plot calculated. At least 100 steps were used per plot. A typical plot is given in Figure 7.15.

The first step in the tuning process was to determine the optimum input loop dimensions for an untuned input circuit. A series of different loops were manufactured, and Q measurements were made to optimise maximum Q while maintaining a minimum offset from the basic cavity resonant frequency. Although detected power

TABLE 7.2

Microwave power levels and test equipment losses

Equipment	Gain / Loss (dB)	Ouput Power (W)
Signal generator		.00016
Cable 1	−2.18	.0001
TWTA	68.1	619.4
Cable 2	−0.13	601.2
Circulator (forward)	−0.15	580.8
Waveguide coupling (forward)	−0.26	547
Cable 3	−0.51	486.4
Thruster return loss	−9.16	59
Waveguide coupling (reverse)	−0.57	46
Circulator (reverse)	−0.08	45.2
Cross coupler	−30	0.045

was primarily used in this process, in later stages, the detector was replaced with a tuning screw and the Q was maximised, using the reflected power characteristic. As the loop dimensions approached optimum, the second step was to use the input tuning screw to minimise the value of the reflected null and increase the value of the detected peak. A clear indication of when optimum loop dimensions were achieved was a set of symmetrical peak and null plots, as shown in Figure 7.15. Asymmetry in these plots indicated that the input circuit and cavity resonant points were not correctly aligned. A further consideration in the loop dimensions was the need to decrease electrical loss in the loop itself due to the very high currents flowing at resonance. The cross section of the loop conductor underwent a number of development cycles to minimise loss and enable stable frequency tracking algorithms to be developed for different input power levels.

Considerable effort was made to maintain accurate calibration of the power levels measured throughout the test programme. A typical set of levels is given in Table 7.2, where the TWTA was at a maximum power output of 620 W. The TWTA gain was the main variable, requiring measurement checks at each test run. Table 7.2 shows the maximum power available at the thruster input was 486 W. It was decided to test the thruster at three nominal power levels: low (150 W), medium (300 W) and high (450 W). These power levels would correspond to one, two and three TWTAs on the satellite while remaining within the specified thermal interface limits.

REFERENCE

1. Shawyer R. Microwave Propulsion-Progress in the EmDrive Programme IAC-08- C4.4.7

8 Flight Thruster Performance Tests

The initial tests quickly showed that some form of frequency tracking would be necessary during test runs. In the classic development sequence for such equipment, open-loop measurements were made before manually stepped algorithms were developed. The end result would be tracking algorithms stored in the processor within the FGCU. The frequency tracking algorithms were necessary because, at initial power turn-on, the input loop rapidly increases in temperature due to high currents at resonance. This results in small dimensional changes and hence, a shift in resonant frequency. The feedback effect of frequency shift gives an inherently unstable circuit at turn-on, and also at any point in the test run when resonance is lost. This effect is countered by determining a frequency offset to be applied at power turn-on, followed by a stepped frequency increase during the following phase of input circuit warm-up. The rate of frequency increase is determined by monitoring the reflected power telemetry at the Telemetry and Control Unit (TCU). The level is maintained above a critical level to avoid loss of resonance. This appears counter-intuitive at first, however the critical level of reflected power ensures that full resonance is not achieved whilst the frequency shift due to the warm up of the input circuit is taking place, which would otherwise cause instability. Once the input circuit has reached a stable temperature with the input frequency at resonance, the effect of the wall temperature increase comes into prominence. The temperature increase results in an effective increase in cavity length, and it becomes necessary to progressively reduce the input frequency to maintain resonance. Once again, reflected power is monitored at the TCU, and the rate of frequency decrease is varied to keep reflected power above the critical level. Figure 8.1 gives a typical frequency tracking result for a medium-power, 300 W test run.

For this test run (FTM 169), the initial frequency offset was 0.2 MHz. After 6 seconds from power on, the frequency was increased for 14 steps of 0.05 MHz at approximately one step per second. This was followed by a 7-second period at constant frequency, followed by a frequency decrease of 7 steps, at one step approximately every 4 seconds. The minimum reflected power telemetry was set to 0.03 V. Table 8.1 gives typical tracking algorithms for the three nominal power levels.

For a flight-qualified engine, the algorithms would be determined by characterising each flight thruster and mapping the data into the control processor of the Frequency Generator and Control Unit (FGCU). Feedback data of detected power, reflected power and baseplate temperature would be used to control the input frequency over the full qualification temperature range for the required input power range.

The force measurement rig is designed to measure the reaction force (R) rather than the Thrust (T) produced by the thruster. As a reminder of the basic theory given

DOI: 10.1201/9781003456759-8

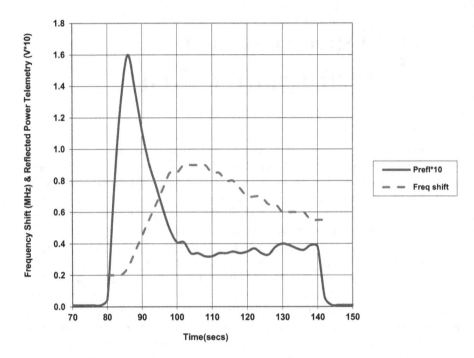

FIGURE 8.1   Frequency tracking for a 300 W test run.

TABLE 8.1
Frequency Tracking Algorithms.

Nominal Input Power	150 W	300 W	450 W
Freq offset (MHz)	0.05	0.2	0.25
Freq step (MHz)	0.025	0.05	0.05
Rate of Freq increase (seconds/step)	2	1	1
No. of Freq increase steps	6	14	20
Period of constant Freq (steps)	15	7	6
Rate of Freq decrease (seconds/step)	5	4	4
No of Freq decrease steps	9	7	5
On period (seconds)	90	60	60
Minimum Prefl. (v)	.005	.03	.07

in Chapter 2, the difference between the two forces is best illustrated by assuming the thruster is in free space, as shown in Figure 8.2.

The net force (F) created within the thruster is given by the basic equation:

$$F = Q\left(F_{g1} - F_{g2}\right)$$

where F_{g1} and F_{g2} are the radiation forces caused by group velocities V_{g1} and V_{g2} at the two ends of the thruster.

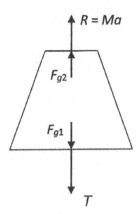

FIGURE 8.2 Thruster force diagram.

This internal force F is measured by an outside observer as the Thrust T, a force acting against the observer in the direction shown.

Newton's laws state that T must be opposed by an equal and opposite reaction force, R, such that

$$R = Ma$$

where $M =$ mass of the thruster
$a =$ acceleration of the thruster in the direction shown.

Clearly, where T and R exist, they will cancel out any attempt to measure them by simply placing the thruster on a balance. This is demonstrated by the results of the calibration test FTM 177, shown in Figure 8.3.

The calibration test runs were carried out to determine the effect of spurious forces generated during test runs and were carried out at the end of the performance test programme. They were carried out with the electronic balance carrying the full weight of the thruster, and therefore no increase in pre-load due to thermal expansion or reaction force was measured. The only force data recorded by the balance was due to spurious forces, EMC effects, or the resolution of the balance itself. Test FTM 176 was with the thrust vector up, and Test FTM 177 was with the thrust vector down. Both tests were carried out at medium input power. The mean spurious force measured over both runs was −0.102 g with a standard deviation of 0.192 g. Note that the measurement resolution of the electronic balance is ± 0.1 g.

The force measurement rig is designed to measure the reaction force dynamically by measuring the change in acceleration of the centre of mass of the thruster caused by the reaction force. During a test run, because there is no thermal compensation in the flight thruster design, the walls of the thruster will expand. The large difference in spring constants between the suspension spring and the electronic balance means this wall expansion will cause the centre of mass of the thruster to move. The movement is recorded as an increase in the pre-load, measured on the electronic balance. The acceleration in this movement, caused by the reaction force, is measured as an

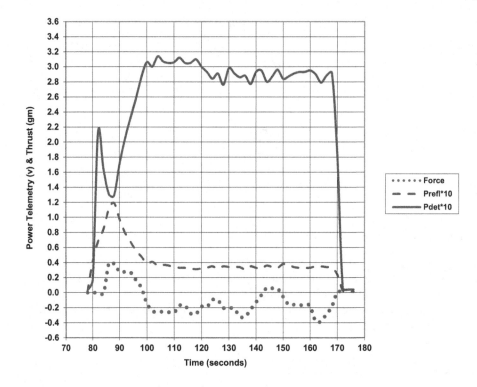

FIGURE 8.3 Calibration test FTM 177 results.

increase or decrease in the pre-load depending on the attitude of the thruster. Thus, with the thrust vector down (as illustrated in Figure 8.2), the reaction force is up, and the pre-load increase will be slightly decreased. The effect is illustrated in the test-run simulation results given in Figure 8.4.

A plot of actual temperature rise is given, while up and down force measurements are simulated to show the effect on a single chart. The reference slope is also shown, which enables the reaction force plots to be calculated. For actual test runs, the reference slope is taken to be a straight line between the start and stop times of the recorded test data. The reference timings account for some of the test data scatter, as the timing resolution is ±2 seconds. The linear reference slope can be seen to be a good approximation to the actual temperature data in Figure 8.4. Test runs, carried out during the development and performance test programmes were designed to produce the following data:

- Reaction force
- Travelling Wave Tube Amplifier (TWTA) output power
- Cavity-detected power
- Cross coupler-reflected power
- Input frequency
- Temperature data at 6 points on the thruster

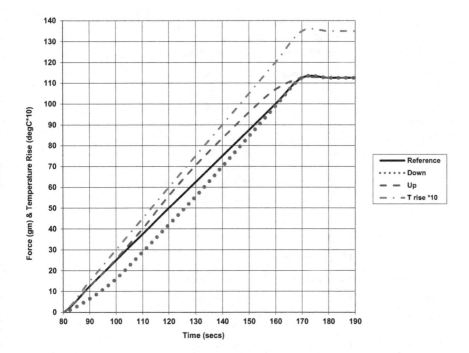

FIGURE 8.4 Simulation of test runs.

All data except the TWTA output power and input frequency were recorded at 2 second intervals on the main computer. TWTA output power was noted with spot measurements throughout the run and was seen to be constant. Frequency data was taken at each frequency step of the tracking algorithm. The development test runs were carried out in a 200-second sequence, with "Power on" periods of 30 seconds, 60 seconds or 90 seconds. The first 15 performance tests were carried out with "Power on" periods of 90 seconds, and the last 4 performance tests had a 60-second "Power on" period. Immediately prior to starting the test run sequence, the resonant frequency was determined. Attenuator 2 was set to −6 dB, and the signal generator output amplitude was set to −30 dBm. Resonance was found by searching for a null using the reflected power readout on the TCU. Attenuator 3 was left at 26 dB. The nominal input power at the thruster for this measurement was 4 W. Attenuator 2 was then increased to 26 dB, and the signal generator amplitude was set to

−15 dB lower power runs
−11 dB medium power runs
−9 dB high-power runs

Note that the TWTA gain variation gave a scatter of actual thruster input powers around the nominal 150 W, 300 W or 450 W. However, the key performance figure for the thruster is the specific thrust in mN/kW over the full, specified power range.

The test sequence is shown in Table 8.2.

Six thermocouples were available for temperature monitoring and were deployed as given in Table 8.3.

TABLE 8.2
Flight Thruster Test Sequence

Time (seconds)	Operation
0	Sequence Timer On
10	Temperature Recorder On
20	Power Recorder On
30	Balance Recorder on
90	Power on
120/150/180	Power Off
200	Telemetry Recorders Off

TABLE 8.3
Thermocouple Positions.

Thermocouple	Position
T1	Thruster baseplate
T2	Thruster top plate
T3	Thruster input
T4	Thruster wall
T5	Free waveguide coupler/Ambient
T6	Thruster detector/cable 3

T1 to T4 were maintained in the same position throughout the test programme. T5 was initially positioned on the static section of the free waveguide coupler. This was to monitor for any change in loss in this component. Once it had been established that the loss was constant, the thermocouple was repositioned to monitor ambient temperature. This enabled any correction to the temperature data due to residual EMC effects to be determined. The maximum correction required was 1.5°C. T6 was initially positioned close to the thruster detector to monitor for any change in detector losses, but as these remained undetectable, the thermocouple was repositioned on the input cable (cable 3), after a failure of this cable. The cable losses were causing very high temperature increases during high power runs. For a 90-second, medium-power run, the temperature increased by 33°C. Following the replacement of the cable, subsequent test runs were restricted to 60-second "power on" periods. Thermocouples T1 and T2 were to provide crucial data for confirming the theory of operation.

The main performance parameter of an EmDrive thruster is the specific thrust. The results of the 19 performance tests, which characterised the specific thrust over the nominal input power range of 150 W to 450 W, are given in Table 8.4. The forward power (Pfwd), given in Table 8.4, is calculated from the calibrated power monitor on the TWTA output and the measured losses up to the thruster input. The reflected power, given in Table 8.4, is calculated at the thruster input, from the calibrated telemetry preferences, and the measured losses are fed to the cross coupler output. The input power to the thruster (Pin) is the difference between the forward and reflected powers.

TABLE 8.4

Flight Thruster Performance Test Results

FTM test number	Thrust vector	Input power	Pfwd (W)	Prfl (W)	Pin (W)	Thrust (g)	Sp thrust (mN/kW)
142	up	medium	292	23	269	10.6	386
143	up	medium	297	24	273	12.3	442
144	down	medium	348	19	329	−8.2	244
145	down	medium	360	13	347	−11.6	327
153	up	medium	366	20	346	10.1	287
154	up	medium	361	25	336	13.1	383
155	up	low	173	13	160	4.5	275
156	up	low	144	14	130	4.3	326
157	up	high	427	71	356	12.6	348
159	up	high	456	59	397	14.2	350
160	up	high	486	59	427	17.7	405
162	up	high	482	59	423	15.7	365
163	down	low	172	12	160	−3.5	215
165	down	high	485	28	457	−17.0	364
167	down	low	162	6	156	−3.2	201
168	down	medium	359	17	342	−8.4	241
169	down	medium	356	32	324	−9.2	278
174	down	high	416	69	347	−12.5	353
175	down	high	395	68	327	−13.5	403

- The mean specific thrust for the 19 tests was calculated as 326 mN/kW, with a standard deviation of 67 mN/kW.
- The mean specific thrust for tests with the thrust vector up was 357 mN/kW.
- The mean specific thrust for tests with the thrust vector down was 292 mN/kW.

The scatter in the specific thrust data is illustrated in Figure 8.5, where the thrust is plotted against input power.

The reaction force for a typical test run is given in Figure 8.6. Test FTM 169 was a medium-power run with the thruster mounted with the thrust vector up. This gives a reaction force down and hence an increase in the pre-load reading on the balance. The actual reaction force plot is therefore positive when the reference slope is subtracted from the electronic balance data.

Figure 8.6 shows that the reaction force follows a positive peaked curve once the Pdet plot reaches a constant at around 100 seconds. At this point, the frequency plot reaches a maximum, showing the input loop temperature has reached a stable value. From this point in the run until power off at 140 seconds, the frequency is stepped down to compensate for the increasing cavity length as the wall temperature increases. The reason why the reaction force plot is a peaked curve while the Pdet plot is constant is because thrust, and hence reaction force, is proportional to the unloaded Q of the cavity, whilst the detected E field is proportional to the loaded Q. The equivalent circuit of the thruster can be considered

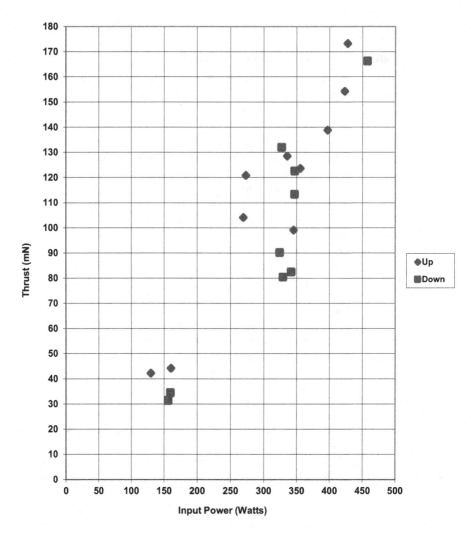

FIGURE 8.5 Flight thruster thrust/power results.

as two resonant circuits in series, one an external circuit consisting of the input and the detector, and one the cavity itself. Circuit theory states that when the impedances of the two circuits match, power transfer will be at its maximum, and reflected power at its minimum. This is the situation at the peak of the reaction force plot, when the input frequency matches the resonant frequency of the thruster. At this point, the Q of the two circuits will both be equal to the unloaded Q of the cavity (Q_u). The Q, as seen at the input of the thruster, will, however, be the loaded Q (Q_l), which will be half Q_u.

Q_l and Q_u are illustrated by Pdet/frequency plots in Figure 8.7. The difference between Q_l and Q_u is shown in Figure 8.8. This difference, when scaled to a maximum of 1.0, is the ratio of thrust to Pdet.

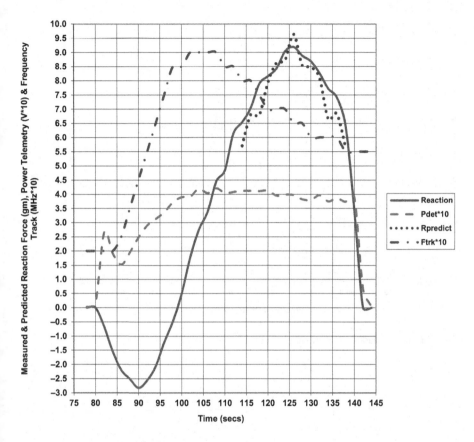

FIGURE 8.6   Test data for run FTM 169.

The reaction force, either side of resonance, was predicted for test FTM 169 using:

- The measured input power
- The mean specific thrust for "up" tests
- The measured wall temperature
- The ambient temperature
- The measured temperature coefficient
- Figure 8.8

The predicted reaction force (Rpredict) shown in Figure 8.6 is in good agreement with the actual measured force around resonance.

The effects of frequency tracking during test runs are illustrated by comparing the results of FTM 157 (Figure 8.9) and FTM 174 (Figure 8.10). Test FTM 157 was a high-power test with the thrust vector down and a "power on" period of 90 seconds. Approximately 50 seconds into the "power on" period, resonance was lost as the cavity thermal expansion was not compensated soon enough by frequency tracking.

FTM 157 shows an increase in Pdet corresponding to the increase in frequency given in the frequency track plot. Pdet reaches a maximum of around 115 seconds.

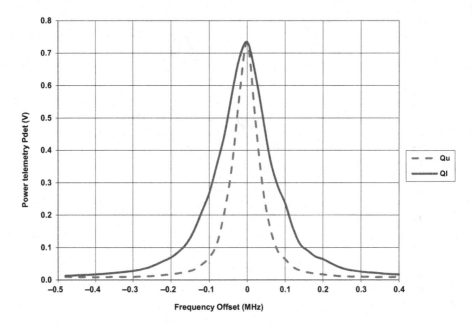

FIGURE 8.7   Loaded and unloaded Q plots.

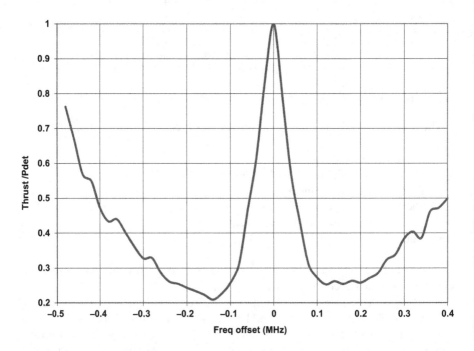

FIGURE 8.8   Plot of Thrust/Pdet ratio versus frequency offset.

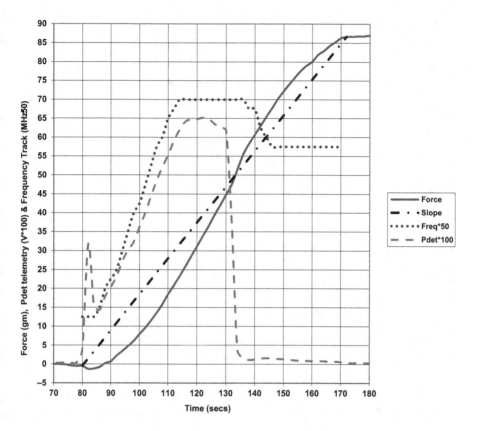

FIGURE 8.9 Test run FTM 157.

The reaction force, given by the difference between the force and the slope plots, has already peaked (around 105 seconds) and is decreasing as loss of resonance approaches. A loss of resonance occurs at 130 seconds and is shown by a sharp drop in Pdet. The subsequent decrease in frequency does not recover resonance due to the feedback effect of a drop in temperature at the input loop, as the high resonance current drops and the reflected power increases.

Test FTM 174 was a high-power test with a "power on" period of 60 seconds and the thrust vector up. Resonance was maintained throughout the run by compensating the cavity's thermal expansion with early negative frequency.

The results of FTM 174 are shown in Figures 8.10 and 8.11. Figure 8.10 shows a frequency increase that leads to maximum Pdet being achieved in 110 seconds. The frequency is then stepped down as the cavity expansion effect takes over from the input circuit warm-up, and resonance is tracked until the power is turned off around 140 seconds later. Note that the criteria for minimum Prefl was 70 mV for test FTM 174, whereas in test FTM 157 it was allowed to drop below 20 mV. The resulting thrust plot for FTM 174 is given in Figure 8.11, together with the Pdet and Prefl plots.

Figure 8.12 shows the temperature data for FTM 154, a medium-power test run. The EMC effect on temperature data can be clearly seen in the ambient results. The highest temperature rise can be seen in the top plate temperature. The high top plate

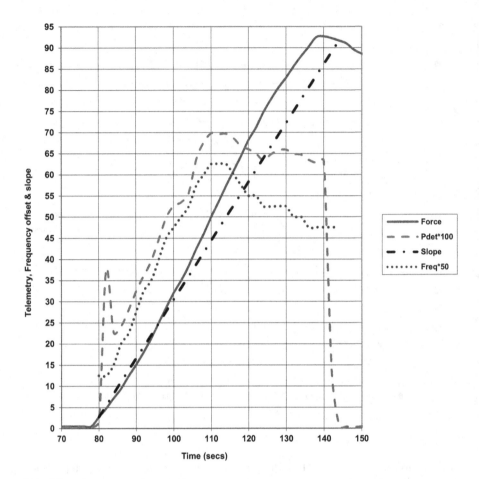

FIGURE 8.10   Test run FTM 174.

temperature compared to the base plate temperature is consistent throughout the test programme. This did not, at first, seem to correlate with theory, as a higher baseplate loss occurs due to the larger baseplate force compared to the top plate. However, by taking into account the differences between the thermal dissipation and radiating areas for both top and bottom plates and using the design factor, from the design software to determine the ratio of losses at each plate, a predicted temperature ratio could be calculated. For the flight thruster, this ratio was calculated to be 0.66. The temperature data for each of the 19 performance tests was then corrected for EMC effects and used to calculate a mean ratio between the measured baseplate and top-plate temperature changes. The mean ratio was 0.69, with a standard deviation of 0.03. This is a very close agreement between the predicted and measured temperature ratios for the baseplate and top plate.

Most importantly, the temperature data provides a clear independent verification of the basic theory by showing that the ratio of power dissipation and therefore force between the two end plates is exactly as the theory predicts.

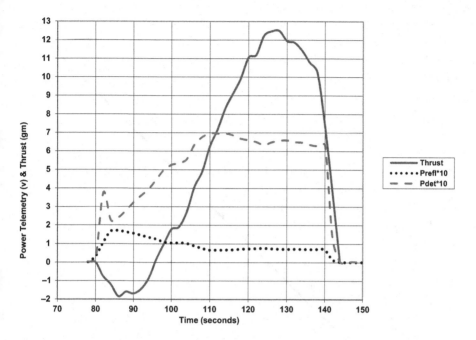

FIGURE 8.11   Thrust plot for FTM 174.

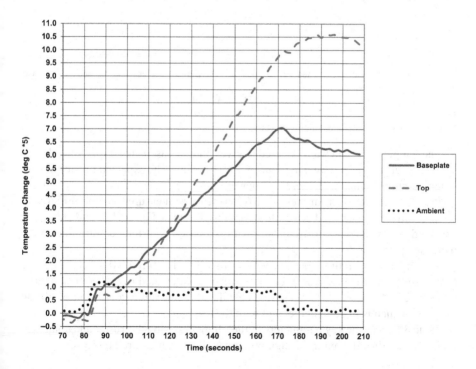

FIGURE 8.12   Temperature data for FTM 154.

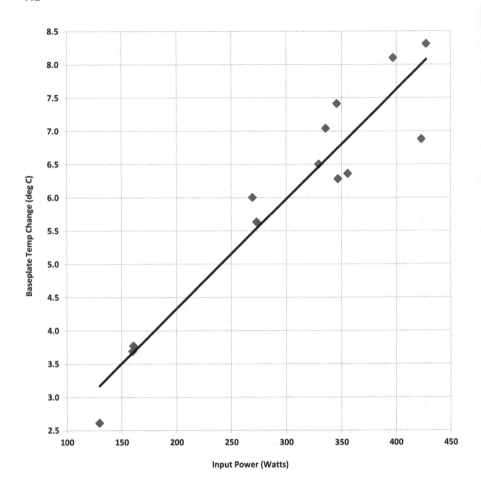

FIGURE 8.13 Baseplate temperature data.

The temperature data was also used to give a plot of baseplate temperature change against input power, as shown in Figure 8.13. This shows the expected linear increase in temperature change with increased power.

Prior to each test run, the resonant frequency was measured for the initial wall temperature of the cavity. The resulting data is plotted in Figure 8.14 and shows the expected linear decrease in frequency with increased wall temperature due to the increased path length.

The mean temperature coefficient from this data is −0.164 MHz/°C. Using the design software, the resonant frequency shift was calculated over the 10° temperature range of the test data. This gave a calculated temperature coefficient of −0.184 MHz/°C, which is in good agreement with the test data.

Table 8.5 gives the key parameters of the Flight Test Model thruster and compares the values predicted from the design process with those measured under test.

Using the Design Software, the resonant frequency at a nominal 20°C was predicted to be 3873.6 MHz from the nominal dimensions. The actual measured resonant frequency was 3850.7 MHz. The lowering of the resonant frequency has been noted on previous thrusters and is attributed to the effect of the input loop, together

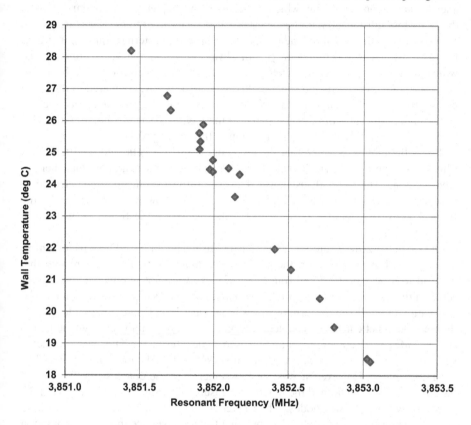

FIGURE 8.14   Cavity temperature and resonant frequency data.

TABLE 8.5
Test results summary.

Parameter	Design	Test
Operating Frequency (MHz)	3,873.6	3,850.7
Max Theoretical Q	73,243	
Unloaded Q		55,172
Specific Thrust (mN/kW)	301	326
End Plate Temperature Ratio	0.66	0.69

with a detailed modification to the nominal geometry of the cavity carried out to increase unloaded Q. The theoretical maximum Q for the cavity is calculated as 73,243, and, as with all previous thrusters, the measured unloaded Q falls well below this at 55,172. The main reason for the discrepancy is the effect of dimensional tolerance on wave-front distortion, which is discussed in Chapter 7. A secondary effect is the slight loss introduced into the cavity by the detector used to measure the Q. This has no effect on the actual unloaded Q, as it is part of the external equivalent circuit. However, the measured unloaded Q will always be below the actual unloaded Q, which gives rise to a lower predicted specific thrust. This is considered to be a contributor to the lower predicted specific thrust of 301 mN/kW, compared to the mean figure given by the test data of 326 mN/kW. Finally, Table 8.5 gives the predicted end plate temperature ratio of 0.66 compared to the measured mean value of 0.69. This data provides a purely thermal confirmation of the design process.

It was therefore concluded that the test programme had validated the Flight Test Thruster design by electrical measurement, mechanical measurement and thermal measurement. The flight thruster test data was fully documented and released to Boeing in September 2010 to complete the contract. Following the expiration of the Boeing NDA, the report was reissued [1] and placed on the SPR Ltd. website in December 2017.

In December 2010, Professor Yang Juan published her first EmDrive paper, unfortunately in Chinese. However, the performance looked sufficiently interesting that we paid to have a professional translation [2]. We were not wrong. The paper presented a theory similar to our EmDrive theory, using the same basic equations. The calculation of predicted performance was based on a 1999 Chinese paper describing the electromagnetic fields in a conical waveguide and used a finite element method to simulate Maxwell's equations for an idealised conical resonator. This led to predicted specific thrusts for a brass cavity of 411 mN/kW for TE011 mode and 456 mN/kW for TE012 mode. However, it was the last sentence in the conclusions that caused most interest, which claimed the practical measurements were 214 mN and 315 mN, respectively, for the two modes. I had been aware that the Chinese were making progress when I was approached by someone from US defence intelligence based in Wiesbaden. I was attending a conference in Toulouse, and he wanted to know what contacts I had received from Professor Yang since my visit to China. Therefore, I knew that something interesting had happened. However, it was not until the second paper was released in 2011 that details of the experimental work were given.

This second paper, finally published in 2012, described the test stand as the "rocket indifferent equilibrium thrust measurement device" [3]. I had seen this thrust measurement rig at NWPU and marvelled at its complexity. It had been developed specifically for testing high-power microwave ion thrusters and could be moved into one of the large thermal vacuum chambers in the laboratory for flight qualification tests. Fundamentally, it was a carefully calibrated rotational balance with a force feedback system. A flexible waveguide was used to transfer microwave power from the fixed to the mobile parts of the balance. Although the force feedback system minimised the waveguide effects, a small load force was maintained, which was calibrated and subtracted from the measured force. The effect of a load force on the thruster was a factor that was seen to be of great importance in subsequent test programmes and will be explained when we address the acceleration and energy aspects of thruster performance in Chapter 10. The direction of the measured force was described as being from the big end to the small end, verifying that it was the reaction force that was being measured. This was the same as the force measured in our flight thruster tests. Two sets of thrust measurements were presented in the paper, which can be accessed on the www.emdrive.com website. The plots show eight test runs where thrust was measured for increasing input power from 80 W to 2.5 kW. The eight test runs are remarkably consistent, showing a stable test regime with a maximum thrust of 720 mN for 2.5 kW input power. To date (July 2023), this is still the highest EmDrive thrust measurement that has been released into the public domain. The total measurement error was given as less than 12%.

The test data gives specific thrust values of 300 mN/kW and 225 mN/kW at the high end of the two power ranges: 300 W to 2,500 W and 80 W to 1,200 W. However, at the lower power values, there is a lot of variation. These discrepancies were attributed to the variation in spectral characteristics of the magnetron being used. Unlike our Demonstrator Engine tests, where the input matching allowed the magnetron to

FIGURE 8.15   NWPU experimental thruster.

pull into the resonant bandwidth of the cavity, this only happened at high powers for the NWPU tests. Spectral characteristics were given for the NWPU magnetron, and these showed a large number of peaks, any of which could have fallen into the resonance bandwidth of the cavity. Nevertheless, the measured specific thrusts at the high magnetron outputs were reasonably close to the 326 mN/kW mean figure from our results.

Not a lot of information was given about the thruster itself, and it was not until a third paper was published in 2014 [4] that more details were given. Figure 8.15 shows a photo of the thruster with its rectangular waveguide cavity input and two stub tuning units. Each end cap can incorporate end plates that are positioned and aligned for maximum Q value. The cavity was operating in TE011 mode at 2.449 GHz, and a maximum Q of 117,495 was claimed, although the matching at this value was almost certainly non-optimum. The maximum theoretical Q was originally claimed to be 62,705. High Q values measured on a network analyser, connected at the cavity input cannot differentiate between the Q of the input circuit and the Q of the cavity itself.

By the time this paper was published, not much information was coming out of China. We knew that SSPAs were being used, and an experiment with a complete engine mounted on an air platform was undertaken. However, our interest was raised when a superconducting thruster was mentioned. Indeed, we were asked if we could quote for the supply of a superconducting thruster, which, after discussion with our UK government contacts, we declined.

REFERENCES

1. Report on the Design, Development and Test of a C-Band Flight Thruster. SPR Ltd September 2010. Issue 2. December 2017.
2. Yang J. et al. Applying method of reference 2 to effectively calculating performance of microwave radiation thruster. *Journal of NWPU* Vol 28, No 6 (2010) 807–813.
3. Yang J et al. *Net Thrust Measurement of Propellantless Microwave Thrusters*. National Natural Sciences Foundation. Chinese Physical Society. 2011. 90716019.
4. Tang Shifeng, Juan Yang et al. Experimental study on microwave resonator tuning systems. 物理学报 *Acta Phys. Sin.* Vol. 63, No. 15 (2014) 154103. *Physics Acta Phys. Sin.* Vol. 63, No. 15 (2014) 154103.

9 The Superconducting Thruster

The basic EmDrive equation for static thrust was derived in Chapter 2 and is given below as a reminder:

$$T = \frac{2PQ}{c}\left\{\frac{V_{g1}}{c} - \frac{V_{g2}}{c}\right\} \qquad (9.1)$$

This equation can be simplified to:

$$T = \frac{2PQDf}{c} \qquad (9.2)$$

where T = Thrust
P = Power
Q = unloaded Q
Df = Design factor

Clearly, thrust is directly proportional to unloaded Q. Actually, when the main performance parameter, Specific Thrust, is considered, Q becomes the dominant variable, as Df tends to remain around 0.9 for a good thruster design. Therefore, it is of great interest to investigate how high the value of Q can become. At room temperature, a typical EmDrive cavity has a Q value of 5×10^4, as demonstrated by the Flight Thruster. The objective of our second-generation (2G) development programme was to substantially increase this value by adopting superconducting technology.

As in any new design, the major parameters must first be established in outline before detailed design can take place. Central to the 2G design process are decisions on which superconducting technology to adopt and at what frequency the thruster will operate. The basic choice of superconducting technology is between a classic liquid helium-cooled niobium cavity or a so-called "high-temperature superconducting" technology. Whilst the classic approach has yielded unloaded Q values of up to 5×10^9 in high-energy physics applications, it was considered that the costs and complexity of liquid helium cooling would be prohibitive. It is worth noting, however, that the internal forces generated during the pulsed operation of these accelerator cavities are so high that they cause the axial length to increase, and compensation techniques using piezo-electric elements must be applied.

High-temperature superconducting technology is available as Yttrium barium copper oxide (YBCO) thin film, thick film or bulk technology for microwave component applications, as well as tape and wire technology for power applications. A literature search and initial design calculations showed that although bulk or thick film technology would be easier to manufacture and is more readily available, it would not

DOI: 10.1201/9781003456759-9

yield the increase in Q that would justify using the superconducting techniques. Thin film technology has, however, caused a decrease in surface resistance values by two orders of magnitude at microwave frequencies. This potentially gives an increase in Q of up to 100 times that achieved to date using copper cavities. However, a major consideration in thin film techniques is the need to coat the film on single crystal substrates. This is a highly specialised process, and a number of companies were approached to identify a possible supplier. A limiting factor of substrate size was quickly identified, with no supplier having experience manufacturing the large thin films envisaged. The microwave design of the thruster therefore went through an iteration phase to establish a frequency and propagation mode that would enable a cavity to be manufactured from the largest substrate available. A supplier, THEVA GmbH in Ismanning, Germany, was then identified who was willing to coat YBCO films and machine sapphire substrates to the necessary sizes.

Having decided on a technology and frequency that appeared feasible, initial mechanical and thermal designs were carried out. It again became clear that there was no precedent for the design approach; indeed, the only applications for super-conducting thin films in the open literature were for microwave integrated circuits, which are tiny by comparison. The mechanical design iterations resulted in a cavity constructed from substrates mounted on Kovar carriers and bonded using a special-ised cryogenic epoxy. The thermal design centred on using a commercially avail-able liquid nitrogen (LN2) Dewar with total loss of the LN2 during testing. The completion of the outline design enabled decisions to be taken on the microwave power source and signal generation equipment. A 40-watt solid-state amplifier was selected, operating over a range of 3.7 GHz to 4.0 GHz. The signal was generated using a standard digital microwave synthesizer operating up to 4 GHz.

The selection of YBCO thin film technology, which was only available on flat substrates at that time, necessitated a move away from circular waveguide cavities to a rectangular geometry. The design software, based on Excel and used in each of the circular cavities built to date, was therefore modified to accommodate rect-angular cavities, and the design iteration process started. The maximum substrate size dictated a higher operating frequency than the 2,450 MHz previously used. A nominal frequency of 3,900 MHz was used as a starting point, and a TE01 propaga-tion mode was selected. Development work on the Demonstrator Engine resulted in the successful implementation of a coaxial input, which was selected for the 2G thruster. Low-temperature, semi-rigid cable assemblies and SMA connectors were selected, but the specifications could not guarantee operation at 77°K, the tempera-ture of LN2. However, both input and detector assemblies were manufactured using this technology after advice was received that it had actually worked successfully at even lower temperatures. The surface resistance (R_S) at a frequency of 3.83 GHz, the operating frequency of the Flight Thruster, is shown in Figure 9.1.

A typical surface resistivity for the YBCO film at 77°K was given by the manufac-turer as 8×10^{-5} ohms at 4 GHz. Using this figure, the theoretical unloaded Q for the thruster is 6.53×10^6, leading to a theoretical specific thrust of 34.6 N/kW. The final design frequency was 3.881 MHz.

The cavity design is summarised in Table 9.1, and the results from the design soft-ware, at a dimensional resolution of 0.5 mm, are given in Figures 9.2 and 9.3.

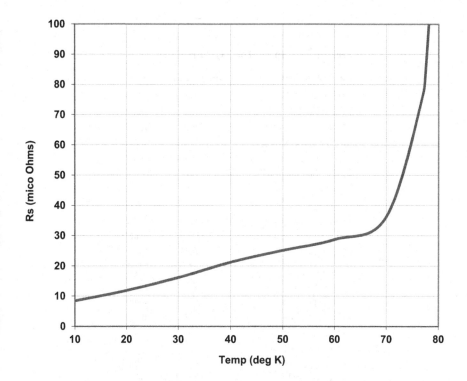

FIGURE 9.1   Yttrium barium copper oxide surface resistance.

As part of the microwave design, spurious modes were identified, and the one of concern was the TE10 mode. The calculated resonant frequency is 3,454 MHz, with an unloaded Q value of 5.89×10^6. As this mode can equally be launched at the input, although with a different wave impedance than the TE01 mode, care was necessary during testing. During the design, the dynamic response of the 2G thruster was mathematically modelled. This analysis was carried out to determine whether control would need to be applied to the input frequency to maintain static thrust under acceleration levels during any proposed dynamic testing. The results showed that, even for the most optimistic surface resistance values, the thrust would be stable up to acceleration levels of 5 g. The effect of acceleration on very high-Q cavities is explored in Chapter 10.

The mechanical design is based on 8 YBCO thin films coated on to sapphire substrates, which are mounted on a carrier assembly to form a cavity resonant at the operating frequency. The carrier assembly consists of six carriers bolted together to form an enclosed cavity, as shown in Figure 9.4.

The carriers are machined from ¼″ (6.35 mm) ASTM-F-15 Kovar plate. Kovar is a special nickel-cobalt-steel alloy with thermal expansion properties that match those of the sapphire substrates that are fixed to the carriers. The substrates are fixed to the carriers using low-temperature epoxy through 10 mm holes in the carriers. The whole thin film carrier assembly is operated in liquid nitrogen at a temperature of 77°K.

TABLE 9.1

Design Summary for 2G Thruster

Cavity Geometry	Rectangular	
Environment	Air	
Dimension a	98.1	mm
Major b Dimension b1	98.7	mm
Minor b Dimension b2	43.7	mm
Taper length L1	144.8	mm
Propagation mode	TE01	
No. of half wavelengths p	3	
Resonant Frequency	3881	MHz
Major guide wavelength	83.9	mm
Minor guide wavelength	164.4	mm
Design factor	0.7935	
Cut-off frequency	3426	MHz
Input position	42.8	mm
Input impedance	332.9	Ohms
Detector position	21.2	mm
Detector impedance	340.9	Ohms
Tuning rod position	65.1	mm
Tuning rod impedance	321.4	Ohms
Cavity material	YBCO thin film	
Theoretical unloaded Q	6,530,163	
Theoretical static thrust	34.571	N/kW
Unloaded 3 dB bandwidth	594	Hz

Hermetic sealing of the cavity, formed by the six carriers, is achieved using a cryo-genic synthetic rubber sealing compound applied to the corner edges and grooves formed by 1 mm chamfers machined on the outer edge of the carriers. An input loop and detector probe are inserted through 6.5 mm holes, and the co-axial flanges are fixed to the carriers using four M2.5 screws. A tuning rod is screwed into an M6 hole. The cavity is attached to the top plate of a 5-L stainless steel cryogenic Dewar using threaded propylene rods and a Dewar fixing frame. The total dry mass of the 2G thruster, including Dewar, was 5.023 kg. The thruster and Dewar are shown in Figure 9.5.

The three important areas of thermal design were matching substrate and carrier thermal coefficients, calculating the temperature rise under power, and determining the rate of LN2 boil-off. The specified thermal expansion coefficient of the sapphire substrate is $5.8 \times 10^{-6} K^{-1}$. After a careful review of the available carrier materials, the nearest match was given by ASTM-15 Kovar with a quoted coefficient of $5.5 \times 10^6 K^{-1}$. Assuming a simple thermal model of dissipation from the end plates only, the tem-perature rise at the substrate/carrier interface was calculated as 1.15°K for the maxi-mum 40 W power input. The amount of LN2 required to chill down the total thruster assembly was calculated at 5.3 L, although this did not account for losses in the fill

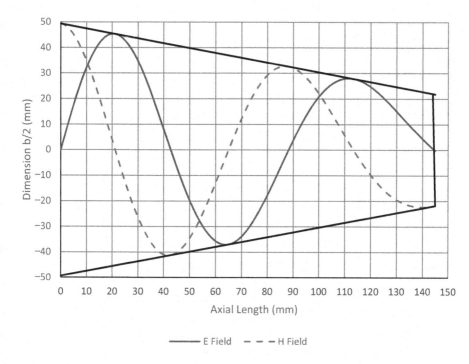

FIGURE 9.2 E and H field phase plots for the second-generation cavity.

procedure. In practice, these losses were significant, and the fill assembly and procedure required a number of changes to enable efficient chill-down to be accomplished. Assuming a 40 W input to the thruster, the rate of LN2 boil-off to maintain 77°K was calculated as 12 g/min. Test results gave a maximum of 3.2 g/min for 40 W input, only slightly above the normal boil-off rate of 2.7 g/min due to heat input from ambient temperature.

An electrical model (ELM) was designed using the modified microwave design software and was used to verify the major design parameters before the commitment to the manufacture of the actual 2G thruster films and carriers was given. The ELM was also used to optimise the design of the input loop and tuner rod, as well as the detector probe. The ELM was fabricated from copper and brass sheet to allow easy modifications to be carried out during the optimisation process and was used to calibrate the 2G test rig, which is shown in Figure 9.6.

The 2G thruster is suspended in its liquid nitrogen-filled Dewar in a compound balance rig. This consists of an angle-iron frame constraining a platform supported on invar rods. The platform supports a 16-kg electronic balance with a resolution of 0.1 g. A balance arm is mounted on a secondary platform, again supported by invar rods. This balance arm, pivoted at one end, rests on the balance pan. The thruster hangs from another invar rod hanging from the other end of the balance arm. At the base of the thruster is an adjuster screw that enables a small offset of the total mass of the thruster to be applied to a second 110 g balance with a resolution of 0.001 g. The compound balance rig provides the capability of measuring low-power Thrust

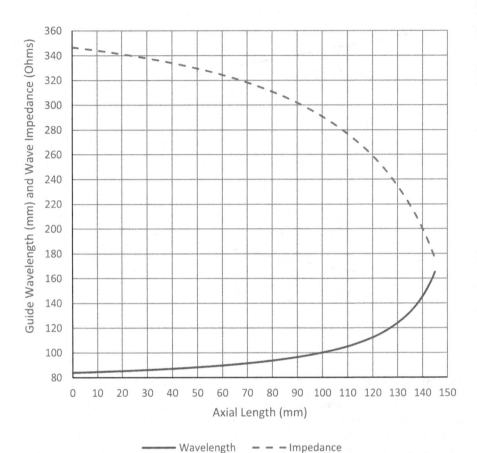

FIGURE 9.3   Guide wavelength and wave impedance plots for the second-generation cavity.

and Reaction Force with high resolution whilst accommodating the full mass of the thruster. Previous experience was gained in the use of a compound balance test rig for static thrust measurements of the Demonstrator Engine. This showed the need to minimise the effects of ambient temperature changes. This led to the requirement to use Invar for the major supporting elements of the rig. The compound balance is housed in a Perspex enclosure with an extractor fan and ducting to remove the nitrogen gas expelled from the thruster. Liquid nitrogen is transferred from a 25-L storage Dewar via a pressurised transfer assembly. Power from the microwave amplifier and a detector signal from the thruster are transferred via flexible coaxial cables. A mathematical model of the compound balance rig was produced and used to check the calibration measurements.

Initial ELM tests were small signal (13 dBm) swept frequency measurements to determine resonant points. The TE01 resonant peak was at 3,834 MHz, with the TE10 resonant peak at 3,426 MHz. A number of tests were carried out to optimise the shaping of the end plate, which had the effect of reducing the number of the higher spurious peaks and increasing the Q of the main peak. The difference in measured

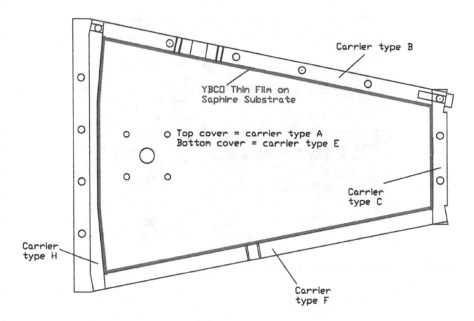

FIGURE 9.4   Second-generation cavity carrier assembly.

FIGURE 9.5   Superconducting thruster cavity and dewar.

Q between no shaping and optimum shaping was a factor of six. A further series of tests were carried out to optimise the detector probe and input loop dimensions. This was followed by initial input tuning carried out with a 3 mm tuning rod to give the resonant peak shown in Figure 9.7.

The exact resonant frequency was 3,778.253 MHz with a Q value of 13,142. The input-return loss was 22 dB. The as-built dimensions were carefully measured, and

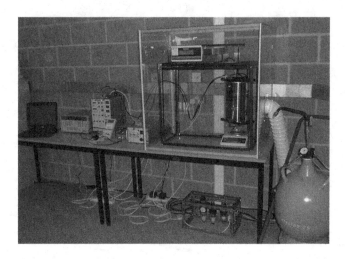

FIGURE 9.6   Second-generation test rig.

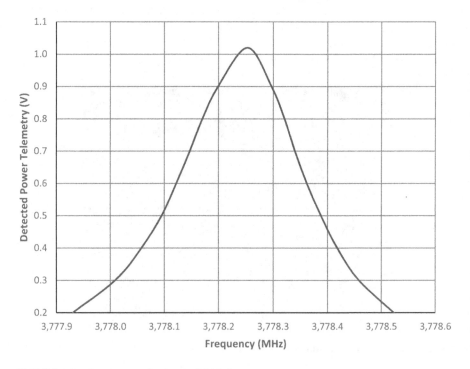

FIGURE 9.7   Resonance plot for the ELM thruster.

the results were used as inputs to the design software. This yielded a predicted resonant frequency of 3,780.15 MHz. The error of .05% is an excellent validation of the
resonant frequency design process. However, the measured Q was well below the
theoretical maximum of 32,672. The reduction represents the difference between
the theoretical unloaded Q of the cavity and the loaded Q, which includes the input

circuit and detector loading. The initial conclusion from the 29 test runs was that the coaxial input and tuning rod enable much higher return losses to be achieved than those obtained with the waveguide input and 3-stub tuner used on the Demonstrator Thruster, but this does appear to be at the expense of some decrease in Q.

The ELM was then used to check out the 40 W power amplifier and associated measurement equipment. Calibration tests and leakage measurements were carried out, with further frequency sweeps confirming that the small signal performance was maintained at 40 W levels. Following the construction of the compound balance test rig, the ELM was used to determine the measurement envelope of the rig. The calibration of the test rig was carried out using standard weights. A number of drift tests were run using different cavity venting techniques to determine thermal effects. The effects of the input and detector co-axial cables were also determined. A series of electromagnetic compatibility tests were also carried out to ensure balance measurements were not subject to electromagnetic interference effects. The compound test rig was then modified to enable compliant and non-compliant suspension of the thrusters. A series of thrust measurements were then carried out, where reaction force using a compliant suspension was measured for the first time. The thruster characteristics for these tests were a resonant frequency of 3781.5 MHz, a Q value of 13,650, and an input return loss of 13 dB. The mean force measurements were a thrust of 61 mg and a Reaction Force of −31 mg. Although the results were low compared to a theoretical force of 270 mg and indeed approached the test rig resolution of 10 mg, the opposite direction of reaction force to thrust was clearly demonstrated, and the design of both the superconducting thruster and the test rig was considered sufficiently validated to go ahead with the expensive procurement of the thin films and carriers.

The fully assembled thruster was initially frequency tested at low power and at room temperature. The results showed a similar profile to the ELM, albeit with a much reduced Q. The first attempt to reach a thruster superconducting condition failed as excessive nitrogen boiled off during the fill process. The thermocouples, used to measure the liquid nitrogen level inside the Dewar, indicated that the required level had not been reached by the time the 25-L storage Dewar was empty. Following a disassembly of the thruster and inspection of the thin films and bonds, the Dewar fill assembly was modified to improve the fill/vent process, and further thermocouples and a level indicator were added. A second cryogenic test was carried out, and 77°K was successfully achieved with enough liquid nitrogen to spare to enable extensive frequency sweeps to be performed. The frequency sweeps gave high-Q versions of the room temperature results, although the exact high-Q resonant point could not be accurately located. A typical sweep is shown in Figure 9.8, where both the TE01 and TE10 peaks are seen, and the increase in Q values with temperature decrease from −177.4°C (95.6°K) to −184.9°C (88.1°K) is clearly shown.

During the subsequent warm-up to room temperature, the YBCO film temperature/resistance curve could be derived from the microwave measurements, and this showed the typical critical point where the cavity had indeed gone superconductive. This is shown in Figure 9.9.

Further modifications were made to the thermocouples to eliminate intermittent failures in temperature monitoring, and a resonance search strategy was prepared.

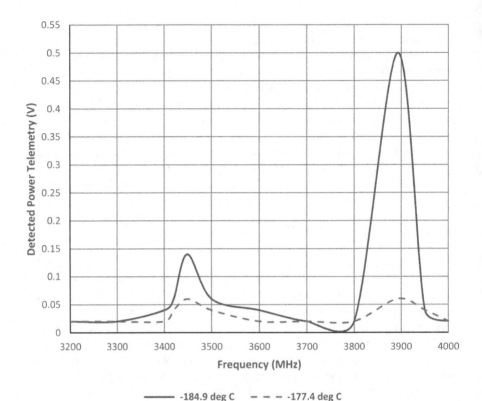

FIGURE 9.8 First cryogenic sweep test.

The third thruster cryogenic test finally resulted in the resonance point being found. This is shown in Figure 9.10. The resulting Q was measured with a frequency sweep using 100 Hz steps. At the resonant frequency of 3,830 MHz (predicted resonance was 3,881 MHz), this represents a resolution of 2.6×10^{-8}. With each step taking 1 second, the frequency search alone represented a formidable task. The measured Q was 6.8×10^6. This is an increase of approximately 500 times the Q measured for the 2G electrical model and is slightly above the theoretical prediction of 6.53×10^6. With the resonant frequency finally established, full power was applied, and the balance readings were monitored with great excitement. However, no force change was detected, and the test quickly ended as the liquid nitrogen ran out.

Following this last test, the thruster was disassembled and inspected. Only two of the substrates remained securely bonded to the carriers, and two films suffered multiple cracks and were no longer usable. The conclusion was that internal stress forces in the films had caused the 0.5-mm sapphire substrates to fracture and the bonds to fail. The field structure within the cavity could no longer be maintained, and therefore no thrust was seen. Whilst this implied that forces had indeed been produced, it clearly meant that a more robust construction would be required. Thicker substrates and pressurised bonding techniques were investigated, but as the funds were rapidly diminishing and the Flight Thruster contract was a priority, the 2G

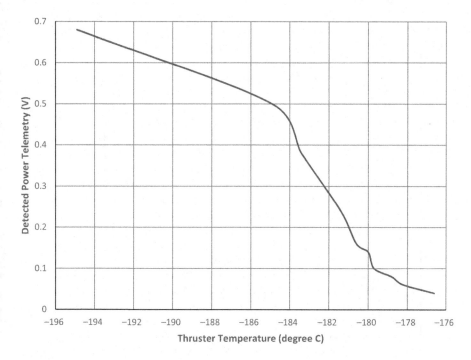

FIGURE 9.9   Second-generation Cavity warm-up test data.

experimental programme was terminated. However, we had shown that a superconducting EmDrive thruster could be constructed using YBCO thin films with electrical performance very close to theoretical predictions, and we had actually achieved
the highest ever reported Q value for a microwave YBCO cavity cooled with liquid
nitrogen.

Around this time, we became aware of some American work on superconducting
EmDrive-type devices. Some experimental work, carried out in January 2011, was
eventually published at a propulsion conference in July 2014 [1]. The results gave a
7 mN force for 10.5 W power at a resonant frequency of 1,047 MHz. A Q value of
1.08×10^7 was measured. The device comprised a circular niobium cavity cooled
to 4.2°K by liquid helium. Although the force was claimed to be generated by the
Lorentz force, created by a differential in radiation pressure exerted by a TM010
wave on the walls of the cavity, it was not clear what created the differential. Upon
closer inspection of the device, it appeared that the input probe at one end of the
cavity was in fact a dielectric component. Thus, at one end of the plate, guide velocity approached c, whilst at the other end, guide velocity in the dielectric was much
reduced. The differential in radiation pressure was being produced by a difference in
guide velocity, completely in accordance with our EmDrive theory.

What had become very clear was that further development of a superconducting
thruster would require a significant increase in funding levels. Therefore, following
the completion of the superconducting experimental work, a number of proposals
were prepared and submitted to different European organisations. We were greatly

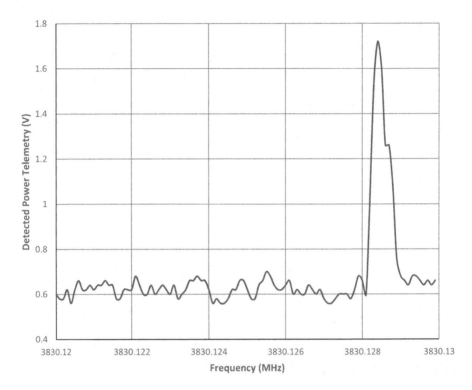

FIGURE 9.10 Second-generation Thruster resonance at 77°K.

assisted by the support of John Prewer, an architect with a strong conviction in the need to work quickly towards solving global problems. At the top of his list was how to make space solar power viable, as every proposal to date had failed due to the enormous cost of launching into orbit the huge solar power satellites that would be needed. The concept of EmDrive providing *"a space elevator without cables"* offered a solution. At the end of 2011, considerable effort was put into a proposal for a "Gamechanger Project" for the Shell Global Energy Company. The project would have taken our superconducting technology to the next level of development, with the objective of testing a liquid hydrogen-cooled, S-band thruster. This was to be a 3-year, multi-million-Euro programme. The eventual applications would range from a hybrid EmDrive/rocket heavy launch vehicle for solar power satellite construction to powering a supertanker. The applications were certainly considered game-changers for the future of the energy industry. We worked with a technical group at the Shell headquarters in Amsterdam to establish the viability of the project, with particular emphasis on the energy balance within the thruster whilst under acceleration.

The programme was initially approved by both Shell and the separate company that had been set up specifically to carry out the work. However, on the morning we were expecting to sign the documents in Amsterdam to start the programme, a dramatic intervention by a high-level American executive immediately cancelled all further work. No clear explanation was given for the cancellation, but it was noted

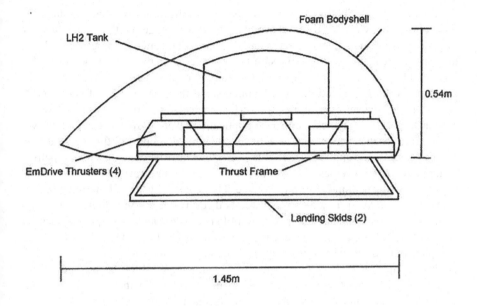

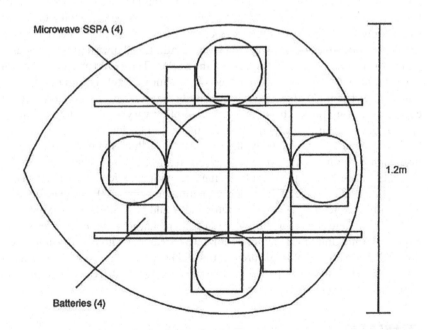

FIGURE 9.11 Second-generation Demonstrator UAV.

that, within weeks, Shell was given permission to drill for oil in Alaska, something that nobody really expected would ever happen. It was suggested that there had been direct US government intervention to halt this particular EmDrive programme. I suspect another reason was simply the significant reduction in the marine fuel oil market that would follow the adoption of EmDrive for tanker applications. Nevertheless, the proposal work highlighted the need to understand the performance of very high Q cavities under acceleration. Whilst the very low acceleration of a supertanker would not cause problems, typical aerospace applications were a different matter. It was considered necessary, however, to demonstrate what a simple 2G thruster could do in a basic aerospace application. To this end, a four-thruster configuration propelling an unmanned aerial vehicle (UAV) was proposed to demonstrate that EmDrive flight is possible using a simple, low-cost vehicle. This UAV is illustrated in Figure 9.11.

The 62.5 kg UAV is based on a lightweight thrust frame carrying four 2G thrusters with their associated solid-state power amplifiers (SSPAs) and batteries. The thruster design is based on a 2.45 GHz cavity operating in TE213 mode. YBCO thin film end plates are used, which give superconducting operation when cooled with liquid hydrogen (LH2). This ensures the high Q value required for the specified thrust output of 463 N/kW at an acceleration of.05 m/s^2. Each SSPA incorporates a microwave signal generator and frequency control system and is powered by a 24-volt, 16-ampere-hour lithium polymer battery. The output section of the SSPA comprises a circulator, a load, and forward and reflected power sensors. The SSPA is rated at 500 W minimum output power and is cooled by the hydrogen gas, which boils off from the cavity cooling system. A 50L, thin-walled, stainless steel tank contains the LH2 supply, which is pressure fed to the cavities via control valves operated by a control system monitoring the cavity temperature. Additional diverter valves are used as part of the temperature control system of the SSPAs. The hydrogen gas is then vented from the top of the UAV. Internal pressure within the tank is limited by a safety relief valve. The flight control system is based on a standard, radio-controlled quadcopter system, giving pitch roll and yaw control by varying the individual thrust of each of the four thrusters. Full telemetry of the propulsion system test data would be incorporated. The UAV body shell is formed of thick foam to provide thermal insulation and limit liquid hydrogen boil-off. It was proposed that initial test flights would be carried out at the manufacturing site, with later demonstration flights covering a maximum range of 1.6 miles at an altitude below 400 feet. All flights were to be within range of sight of the operator and would be limited to a maximum flight time of 12 minutes before a refill with LH2 and recharging of the batteries would be required. A maximum velocity of 11 mph was predicted with a maximum acceleration of.005 g. Although this UK programme did not proceed, we understood that both the USA and China were pursuing superconducting EmDrive research with UAV applications in mind.

REFERENCE

1. Fetta GP, Cannae LLC. Numerical and Experimental Results for a Novel Propulsion Technology Requiring no On-Board Propellant. 50th AIAA/ASME/SAE/ASEE Joint Propulsion Conference. July 2014.

10 The Trouble with Acceleration

One of the questions always asked about EmDrive is how it complies with the conservation of energy. Once the results of mission studies are scrutinised, it can appear as if the energy balance results in an over-unity solution. If a very simple approach is followed, the kinetic energy transferred to a vehicle seems to end up higher than the electrical energy put into the vehicle. Clearly, this cannot be right, and the explanation lies in the fact that the electrical energy is input over incremental increases in velocity. It is vital that the thruster does not lose thrust during this acceleration, and a mechanism involving an internal Doppler shift in the travelling wave inside the cavity is examined in this chapter, and an equation for thruster efficiency is derived.

The Doppler shift that occurs in a high-Q cavity during acceleration is best described using the following simple mathematical model, first published at the IAC-13 conference held in Beijing [1]. The cavity is shown in Figure 10.1.

The cavity is supplied with a microwave frequency of F_0, the resonant frequency of the cavity. Assume the EM wavefront propagates initially from the large end plate towards the small end plate. At the end of this forward transit, the wavefront is reflected at the small end plate. At this time, due to cavity acceleration, the cavity velocity has increased to V_r, whereas the wavefront has a constant guide velocity of V_{g2}. The relative addition of these velocities gives the reflected wavefront a Doppler shift, resulting in a reduced frequency F_r for the reverse transit. On reaching the large end plate, the wavefront is again reflected and subjected to a second Doppler shift, resulting in the forward frequency F_f. The increase in frequency is calculated from the relative addition of the guide velocity V_{g1} and the new cavity velocity, V_f. The sequence of Doppler shifts at each reflection of the wavefront will continue as the stored energy in the cavity builds up. If the cavity acceleration A is zero, then the relative velocity between the large and small plates, at the time of wavefront reflection is also zero. This will result in an overall zero-Doppler shift. However, with a positive acceleration, the overall Doppler shift will be negative. With a negative acceleration, the overall Doppler shift will be positive.

The transit time t_s (seconds) for a wavefront to travel from one end of the cavity to the other is given by:

$$t_S = \frac{p}{2F_0} \qquad (10.1)$$

where p = number of half wavelengths
F_0 = Resonant frequency (Hz)

DOI: 10.1201/9781003456759-10

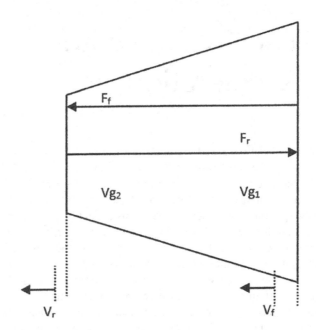

Cavity Acceleration A

FIGURE 10.1 Cavity under acceleration

The cavity acceleration A (m/s/s) used in the model is given by:

$$A = \frac{QuAs}{50} \tag{10.2}$$

where $As =$ acceleration to be simulated.

$Qu =$ unloaded Q

Clearly, a full simulation would require Q_u forward and reverse transits, but as this would be impractical to model, a reduced number of transits is used, with the necessary increase in model acceleration A.

For a forward transit, the velocity at reflection, V_f (m/s), is given by

$$V_f = V_r + At_s \tag{10.3}$$

The reflected frequency F_f (Hz) is given by:

$$F_f = F_1 \left[\frac{1 + \dfrac{V_f}{V_{g1}}}{1 - \dfrac{V_f}{V_{g1}}} \right] \tag{10.4}$$

For a reverse transit, the velocity at reflection, V_r (m/s), is given by:

$$V_r = V_f + At_s \tag{10.5}$$

The Reflected Frequency F_r (Hz) is given by:

$$F_r = \left[\frac{1 - \dfrac{V_r}{V_{g2}}}{1 + \dfrac{V_r}{V_{g2}}} \right] \tag{10.6}$$

The model solves the reflected frequency equations for 50 simulated forward and reverse transients to give $F_{f(n=50)}$. The total Doppler shift F_D (Hz) is given by:

$$F_D = F_{f(n=50)} - F_0 \tag{10.7}$$

The 3 dB bandwidth B (Hz) of the resonant cavity is given by:

$$B = \frac{F_0}{Qu} \tag{10.8}$$

At the point where the total Doppler shift reaches $B/2$, the stored energy and hence thrust fall to half the static thrust.

The model was used to predict the specific thrust as the total Doppler shift increased for different cavity designs.

The dynamic performance of the flight thruster described in Chapter 7 was modelled with a cavity $Qu = 50,000$, $F_0 = 3.85$ GHz and an input power of 1 kW. The results are shown in Figure 10.2.

Figure 10.2 shows that for such a low Q factor, spacecraft acceleration levels up to 1 m/s/s cause negligible Doppler shifts (<1 Hz) such that no loss of Q is experienced and therefore the thrust remains constant. To illustrate typical applications, the spacecraft mass is also plotted against acceleration, assuming 1 kW of microwave input power to the thruster. It can therefore be concluded that for first-generation, in-orbit propulsion applications, thrust will be constant throughout the total thrust period and equal to the measured static thrust.

A superconducting thruster, similar to the one described in Chapter 9, was then modelled and cooled first with liquid nitrogen and then with liquid hydrogen. With the thruster cooled to 77 K using liquid nitrogen, a Q of 3.7×10^6 was assumed. The specific thrust from the model is shown in Figure 10.3.

Figure 10.3 shows a static-specific thrust of 16.5 N/kW. With this modest value of Q, it requires high acceleration to cause a significant reduction in specific thrust. In this case, an acceleration of 1000 m/s/s (100 g) gives a specific thrust reduced to 4 N/kW.

Figure 10.4 gives the results for a lower frequency cavity cooled by liquid hydrogen, thus operating at a temperature of 20 K and achieving an unloaded Q of 3.9×10^7. A static specific thrust of 173 N/kW was predicted, and the reduction of specific

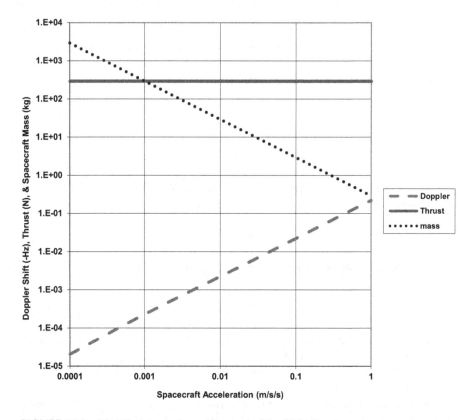

FIGURE 10.2   Modelled dynamic performance of the flight thruster.

thrust with acceleration is more pronounced, giving a specific thrust of 11 N/kW for an acceleration of 20 m/s/s (2 g) with no compensation.

A compensated version of this cavity was designed and modelled, where the axial length of the cavity was modified according to the acceleration experienced by the thruster. The cavity extension for a positive acceleration of 20 m/s/s is illustrated in Figure 10.5. The extension results from a pulsed voltage being applied to piezo-electric elements in the sidewall of the cavity. The pulse length is determined by the time constant of the resonant cavity. Clearly, this simple form of compensation cannot completely compensate for the Doppler shift throughout a full pulse cycle, but Figure 10.4 shows that the specific thrust at 20 m/s/s can be improved to 92 N/kW.

A large, high-power thruster was then designed, operating at 900 MHz. This thruster again used an YBCO superconducting coating and was cooled with liquid hydrogen. The compensation technique included both cavity length extension and frequency offset, with a lower duty cycle than the 3.85 GHz thruster. A specific thrust of 9.92 kN/kW was predicted with an acceleration limit of 0.5 m/s/s. The design became the basis for a number of vehicle designs employing hybrid propulsion systems. One such vehicle was described in a paper presented at a Council for European Studies (CES) conference in 2009 [2]. The basic concept of a hybrid spaceplane was a vertical take-off and landing carrier vehicle using eight EmDrive lift engines, two

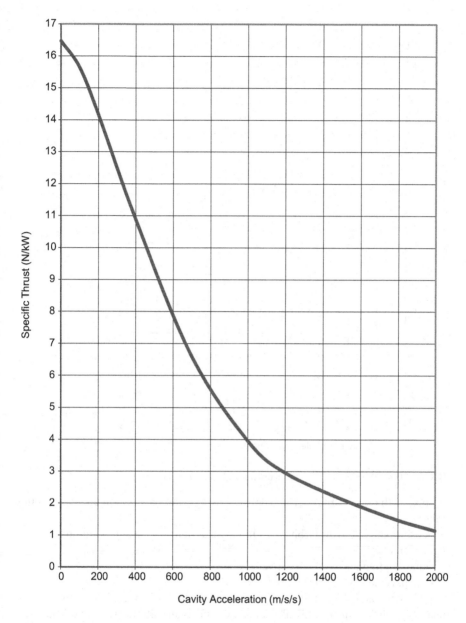

FIGURE 10.3   Modelled specific thrust using LN2 cooling.

hydrogen-fuelled jet engines with vertical lift deflectors, and up to six hydrogen-/ oxygen-fuelled rocket engines. Electrical power would be provided by two fuel cells run on the boiled-off hydrogen gas from the lift engines and liquid oxygen. The hybrid spaceplane was designed to carry a variety of payloads, and analyses were carried out for a number of missions, including:

Long-distance passenger transport using suborbital flight.

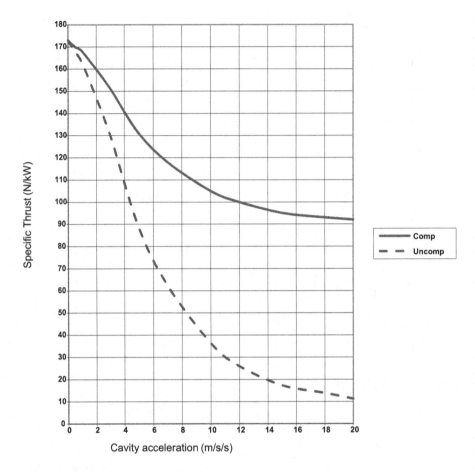

FIGURE 10.4 Modelled specific thrust for an LH2-cooled thruster

Low Earth Orbit (LEO) payload delivery using additional orbital engines and fuel tanks.

Geostationary Earth Orbit (GEO) payload delivery using additional orbital engines and fuel tanks.

Lunar landing and payload delivery.

The spaceplane is illustrated in Figure 10.6.

A further spaceplane design was presented at the IAC-14 conference in Toronto and then published in 2015 [3]. The launch vehicle is an "all-electric" single-stage-to-orbit (SSTO) spaceplane using a 900 MHz, eight-cavity, fully gimballed lift engine. A 1.5 GHz fixed orbital engine provides the horizontal velocity component. Both engines use total-loss liquid hydrogen cooling. Electrical power is provided by fuel cells, fed with gaseous hydrogen from the cooling system and liquid oxygen (LOX). A 2-tonne payload, externally mounted, can be flown to low Earth orbit in 27 minutes. The total launch mass is 10 tonnes, with an airframe styled on the X-37B that allows aerobraking and a glide approach and landing. An outline diagram of the SSTO spaceplane is given in Figure 10.7.

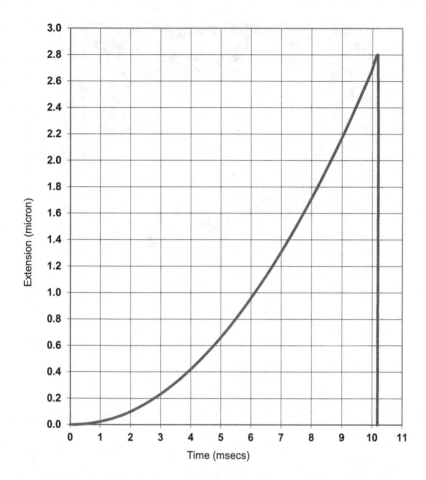

FIGURE 10.5 Cavity extension for 20 m/s/s acceleration.

This shows the lift engine centrally mounted above the C of G of the vehicle, with three small, C-band 2-axis gimballed EmDrive thrusters to give roll pitch and yaw control. The bulk of the volume is taken up with the two liquid hydrogen tanks, which provide 21,000 L of LH2. The LH2 provides cooling for the EmDrive engines and gaseous H2 for the fuel cells. The liquid oxygen tanks provide 260 L of LOX for the fuel cells, rated at a nominal 500 kW. The 2,000 kg payload is mounted on rails on the top surface of the space plane. A launch mass of 9,860 kg was estimated, including the payload, with a vehicle dry mass of 6,080 kg.

Although these designs incorporated Doppler compensation, acceleration levels were still limited by the use of axial length and frequency compensation carried out over the time constant of the cavity. The next step in solving the acceleration problem came after consideration of the energy balances of both SSTO missions and the proposed Interstellar Probe, described in the same paper. A useful method of reviewing the results of EmDrive-propelled missions is to calculate the total energy balance of the acceleration phases of the missions. This enables the energy efficiency to be

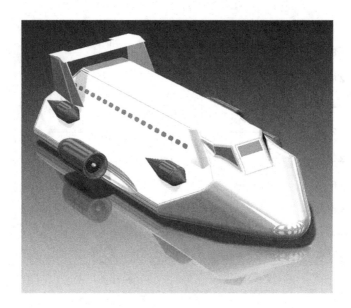

FIGURE 10.6   Hybrid spaceplane.

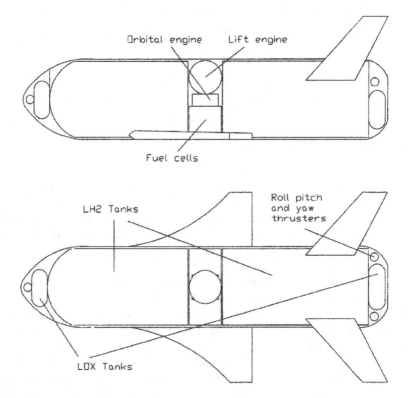

FIGURE 10.7   Outline diagram of the SSTO spaceplane.

calculated in terms of the total kinetic energy of the spacecraft at terminal velocity divided by the total energy input during the acceleration period. The overall energy flow is illustrated in Figure 10.8.

From equation (9.2) EmDrive thrust is given by:

$$T = \frac{2PQDf}{c}$$

The instantaneous mechanical power produced by thrust T is given by:

$$P_{mech} = TV_1 \qquad (10.9)$$

where V_1 = spacecraft velocity

Over the total acceleration period, the total energy input to the spacecraft is given by:

$$E_{in} = TV_{av}.t_a \qquad (10.10)$$

where V_{av} = average spacecraft velocity
    t_a = acceleration period

Substituting equation (9.2) in equation (10.10)

$$E_{in} = \frac{2PQDfV_{av} \cdot t_a}{c} \qquad (10.11)$$

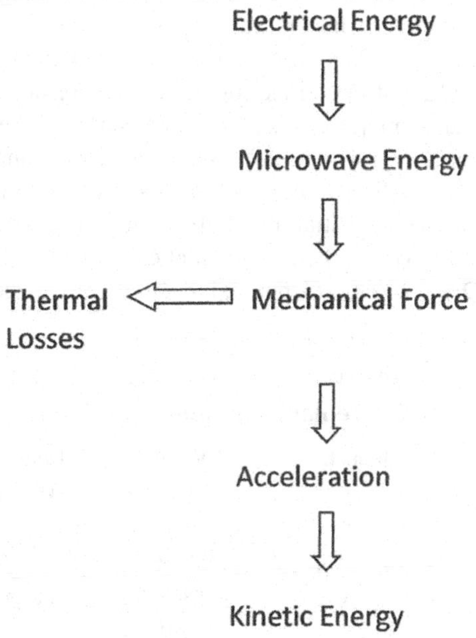

Electrical Energy

Microwave Energy

Thermal ⟸ **Mechanical Force**
Losses

Acceleration

Kinetic Energy

FIGURE 10.8 Diagram of energy flow.

The kinetic energy of the spacecraft at terminal velocity V_T is given by:

$$E_k = \frac{MV_T^2}{2} \tag{10.12}$$

where M = mass of spacecraft
V_T = terminal velocity

The efficiency of the EmDrive thruster can therefore be calculated from:

$$e_t = \frac{E_k}{E_{in}} \tag{10.13}$$

This equation was used to calculate the efficiency of the orbital engine in raising the spaceplane velocity to orbital velocity at LEO.

The in-orbit SSTO spaceplane parameters are given in Table 10.1.

From equation (10.11):
$E_{in} = 6.75 \times 10^{11}$ J
From equation (10.12)
$E_k = 2.45 \times 10^{11}$ J
Therefore, from equation (10.13), mission efficiency for the orbital thruster
$e_t = 0.363$
For overall engine efficiency in this mission, the electrical power input from the fuel cells should be used, reducing the mission efficiency to
$e_m = 0.243$

This analysis demonstrates that the efficiency of an EmDrive engine is dependent on both engine parameters and the mission in which it is used.

However, the conference papers had apparently created interest in methods of solving the acceleration problem in both the USA and China. In January 2014, I met with John Fry, the CEO of Fry's Electronics, a large consumer electronics company that at one time had 34 major stores across nine states in the USA. John had flown into London on one of his own aircraft, as billionaires do, and wanted to discuss the math behind EmDrive. I discovered that John was a first-rate mathematician, and the founder of the American Institute of Mathematics, who worked with the National Security Agency (NSA) via their Mathematical Sciences Programme. The basic math behind EmDrive was not disputed; indeed, it was considered trivial. However,

TABLE 10.1
Spaceplane Parameters

Parameter	Value	Units
P	324	kW
Q	8×10^7	
D_f	0.7706	
V_T	7800	m/s
T_a	1300	secs
M	8059	kg

we spent a full day discussing the details of the effects of acceleration on EmDrive. John put forward concepts involving real-time correction of the frequency using very fast digital techniques and assured me that this was of great interest to the US government. It apparently was a topic of discussion with a past US secretary of state on a local golf course. I was very impressed.

I was then stunned to receive an email the next day from Professor Yang Juan, asking me whether I had tried circular polarisation for the propagation mode inside an EmDrive cavity. The coincidence of these two events led me to work out how circular polarisation, and fast digital control of the input phase and frequency would solve the acceleration problem. As is well known to radar designers, the use of circular polarisation enables the separation between the transmitted and received signals. This is because as the signal is reflected from the target, the polarisation is reversed. Thus, by using a radar antenna with two helical elements with reversed polarisations, the low-power received signal can be separated from the high-power transmitted signal, even if they are of the same frequency. This technique can be applied to a third-generation (3G) superconducting cavity illustrated in Figure 10.9.

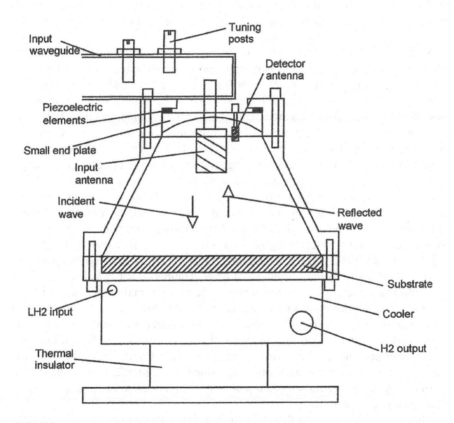

FIGURE 10.9   3G superconducting cavity.

In Figure 10.9, the thruster comprises a shaped small end plate fixed by screws to a taper section, which is fixed to a flat, large end plate. This is the basic cavity shape described in Chapter 2 and shown in Figure 2.8. A single-crystal sapphire substrate is attached to the large end plate. The inner surface of the sapphire substrate, together with the inner surfaces of the taper section and small end plate, are coated with a thin film of YBCO. A liquid hydrogen (LH2) cooler is fixed to the large end plate. After passing through the cooler, the LH2 becomes gaseous due to the input of heat dissipated at the YBCO thin film and substrate. The latent heat of evaporation of the LH2 provides the cooling effect at the YBCO surface, maintaining the film temperature below its critical temperature, thus maintaining its superconducting properties. The cold hydrogen gas (H2) then exits from the cooler and is used to cool the solid-state power amplifier and fuel cell, which are used to power the thruster. Some of the expended H2 is then used in the fuel cell, together with oxygen, to provide electrical power to the solid-state power amplifier. The cooler is fixed to a thrust plate via a thermal insulator. The small end plate assembly contains a shaped section that slides within the fixed end plate. The shaped section and the fixed end plate are separated by piezoelectric elements, which control the axial length of the cavity according to the common pulsed electrical signal applied to them. These piezoelectric elements can also be used to control the precise alignment of the end plates by varying the DC signal level for individual elements.

Microwave power is transferred to the cavity via a waveguide input section. This waveguide section contains two tuning posts, whose length can be adjusted to give the correct impedance match to the microwave source and ensure maximum power transfer from the source to the cavity. The microwave power is transferred from the input waveguide, via an input probe, to a helical input antenna. This input antenna, manufactured from YBCO tape, propagates the microwave power as a circularly polarised electromagnetic wave. This incident wave is reflected from the large end plate. The reflected electromagnetic wave has the opposite polarisation to the incident wave and is detected by a small helical detector antenna, which has a helix geometry that is opposite to that of the input antenna. The detector antenna is designed so that only a very small fraction of the reflected electromagnetic wave is extracted from the cavity, and the polarisation difference with the input waveform ensures that the detected signal level is above any noise signal caused by the input electromagnetic wave. The axial length of the cavity is tuned by the piezoelectric elements such that it is always a whole number of half wavelengths of the input electromagnetic wave. In this manner, the cavity is maintained at resonance, and the input and reflected waves continue to be reflected backwards and forward between the small and large end plates. A critical element of achieving a high-Q cavity is the geometry necessary to ensure that the backward and forward transits of the electromagnetic waves traverse the same path length, independent of the radial position along the wavefront, as described in Chapter 2. The geometry that is necessary to achieve this constant path length, independent of radial position, is illustrated in Figure 10.10.

The taper section ensures that the diameter FH is smaller than the diameter EJ, which gives a projected apparent origin of the electromagnetic wave at position O. The shape of the minor end plate (curve FAH) is designed to ensure that the outer and axial path lengths EF, BA and JH are equal. In addition, any path length, represented

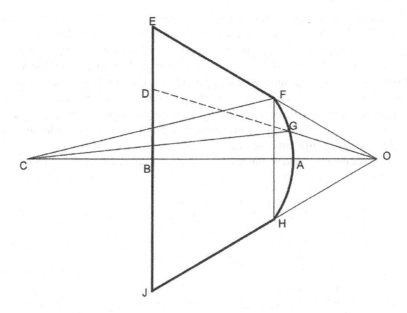

FIGURE 10.10 Cavity geometry.

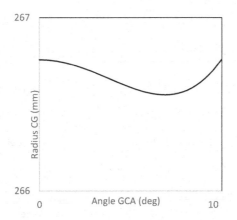

FIGURE 10.11 Typical curve shape for small end plate.

by DG in Figure 10.10, must also be equal to the outer and axial path lengths. This geometry is ensured by calculating the value of the machining radius CG of the curve FAH for any angle represented by GCA. This calculation is carried out by a numerical analysis in which the machining radius CG is iterated for steps in the angle GCA until the path length DG is equal to the outer and axial path lengths EF, BA and JH. The resulting curve shape, FGA, is shown in Figure 10.11, where a typical result of such an analysis is given.

A mirror image of this curve gives the curve AH, and thus the complete concave shape of the small end plate can be machined. The high precision needed to produce the modified spherical shape can be deduced from the vertical scale of Figure 10.11.

As has been explained, when the cavity is subjected to an acceleration, a Doppler shift will occur in the incident and reflected electromagnetic waves. Because the guide velocities are different at the large and small end plates, these Doppler shifts will not cancel each other out. It is therefore necessary to introduce a closed-loop control system to modify the input frequency and phase and provide the input signals to the piezoelectric elements to correct the Doppler shift. Clearly, the correction to the incident wave can only be done at the instant of propagation. Once propagated, the frequency and phase cannot be further corrected. However, the next cycle can be corrected as the phase difference between incident and reflected waves is monitored by the detector antenna. A block diagram of the control system is given in Figure 10.12.

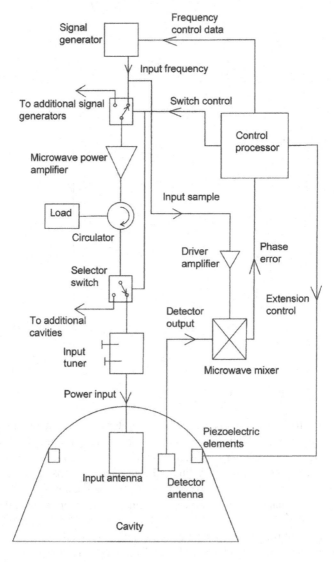

FIGURE 10.12 Control system block diagram.

Figure 10.12 shows the input frequency is generated at a low-power level by the signal generator, a fast digital synthesiser. The frequency signal from the signal generator is sent to a microwave power amplifier via a switch, which can select a frequency signal from any one of a number of additional signal generators to provide redundancy in this critical function. The output of the microwave power amplifier is fed through a standard circulator and load protection circuit to a second selector switch, which enables the output power to be sent to a second cavity in sequence. The microwave power is then fed to the input antenna inside the cavity via an input tuner. The oppositely polarised detector antenna then provides a very low-level fraction of the reflected electromagnetic wave to the input port of a microwave mixer. The local oscillator port of the mixer is fed via a drive amplifier, whose input is a sample of the input frequency being generated by the signal generator. The output of the microwave mixer will therefore have a phase error, corresponding to the Doppler shift difference between the input and the reflected electromagnetic waves. This phase error is then fed to the control processor, where it is processed to produce the frequency control data, which is sent to the signal generator. Thus, a phase-locked control loop is set up to maintain the Doppler shift difference to a minimum under acceleration conditions and thus maintain the high Q of the cavity. The control processor also provides a voltage to the piezoelectric elements to control the extension of the cavity's axial length. However, the input frequency cannot be continuously corrected during constant acceleration, and the accompanying extension of the axial length of the cavity cannot be unlimited. Therefore, the Doppler correction function is carried out over a specific period that starts and stops the power input to the cavity, as shown in Figure 10.13.

Figure 10.13 shows the input power pulse, the Doppler frequency shift, and the cavity length extension for one cavity of a two-cavity engine. In this example, the input power pulse lasts for one second, and the acceleration causes the Doppler frequency to lower the input frequency, according to a curve that can be calculated from

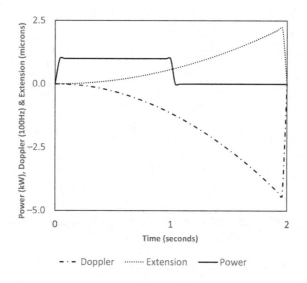

FIGURE 10.13   Doppler correction over two pulse periods.

a numerical analysis of the dynamic response of the cavity. The extension of the cavity axial length therefore increases the cavity length, according to a curve, which is an inverse shape compared to the Doppler curve. At the end of the power pulse, the Doppler shift and extension curves continue as the stored energy and the electromagnetic waves forming that energy approach zero. The thrust output for each cavity is shown in Figure 10.14.

Figure 10.14 shows that during the power pulse to cavity 1, the thrust builds up to the rated thrust output (1 on the vertical axis of Figure 6), in an exponential curve. When the power pulse is switched to cavity 2 in one second, the thrust in cavity 1 falls exponentially to approach zero at two seconds. At the second point, the extension is reverted to zero and the thrust drops to zero, as the cavity is no longer tuned to the Doppler-shifted frequency. Meanwhile, in the period of one second to two seconds, the power pulse is applied to cavity 2, and the thrust from cavity 2 rises exponentially. The cycle continues such that the total thrust remains approximately constant, with small dips each time the extension of a cavity reverts to zero. The thruster was described in a patent [4] published in October 2016 and eventually granted in August 2021.

This 3G thruster design with Doppler compensation gave much improved performance for very high-Q cavities. A number of mission analyses were carried out, allowing acceleration levels of up to 0.1 m/s², which seems incredibly low when compared to typical rocket-propelled missions but nevertheless allowed a proposed flight to the Moon in 37 hours. A 2019 IAC paper [5] presented the results of this analysis, showing the flight divided into three phases. The initial 13-hour acceleration phase gave a gentle flight through Earth's atmosphere, with a maximum velocity of only 70 mph and thus very low drag. This was followed by a cruise of 11 hours at a velocity

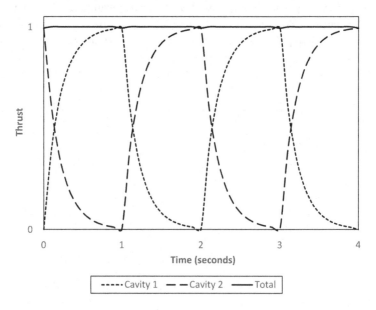

FIGURE 10.14   Thrust output for a two-cavity engine.

of 10,160 mph, which was the terminal velocity of the acceleration phase. The third phase was a 13-hour deceleration at a maximum rate of $-0.1\,\text{m/s}^2$. This remarkable 10-tonne vehicle would transport three astronauts to the Moon and back, with controlled vertical landings on both surfaces of the Moon and Earth. The drama-free flight plan was considered the ideal basis for an eventual space tourism business.

However, by this time, information had become public that indicated that someone was actually flying vehicles with apparent propellant-less propulsion and achieving very high acceleration levels. The gradual leakage of information was eventually confirmed by a US intelligence report, published in June 2021 [6]. The report included the following conclusions:

Most of the UAP reported probably do represent physical objects given that a majority of UAP were registered across multiple sensors, to include radar, infrared, electro-optical, weapon seekers, and visual observation.

And a handful of unidentified aerial phenomena (UAP) appear to demonstrate advanced technology.

The possibility that these vehicles, termed UAP, could actually be experimental drones propelled by EmDrive thrusters had already been suggested in the US in an April 2021 article in the Combat Aircraft Journal [7]. This publication is highly respected by the US armed forces. Indeed, the main observables credited to these UAPs, including operation in multiple environments, namely air, space and underwater, would be readily achieved using EmDrive propulsion. The lack of exhaust signature, the cold vehicle signature, and the detection of RF leakage all corresponded to 3G Emdrive. What was extraordinary, however, was the high speed and very high acceleration levels observed. Apparent supersonic speed, without the associated sonic boom, is usually associated with electronic warfare (EW). Numerous methods, such as range gate stealing, a classic technique described in EW textbooks [8], could be put forward as a possible explanation. However, I was left with the uncomfortable feeling that someone was way ahead of my understanding. Accordingly, I determined to think of a technique that would unlock much higher acceleration levels for EmDrive.

The answer was surprisingly simple. The acceleration limitation is basically due to the law of conservation of energy. Acceleration results in a gain in kinetic energy ΔKE, giving a reduction in frequency due to the Doppler shift and a reduction in stored energy, which results in thrust reduction. To increase acceleration, ΔKE must be limited by reducing the pulse length of the applied power. The problem can be analysed by deriving an equation for the efficiency of the system but restricting the acceleration period to the pulse period rather than the overall acceleration period of the mission, as was used to derive equations (10.11), (10.12) and (10.13).

For any vehicle, the kinetic energy gained by the vehicle in the acceleration period is:

$$\Delta KE = \frac{M}{2}\left(V_2^{\,2} - V_1^{\,2}\right) \tag{10.14}$$

where M is the vehicle mass, V_1 is the initial velocity and V_2 is the final velocity.

The mechanical energy input to the vehicle during this period is:

$$E_m = Fd = FV_a t = \frac{KPt(V_2 + V_1)}{2} \qquad (10.15)$$

where F is the Force, d is the distance travelled, V_a is average velocity, t is the time period, K is the specific thrust and P is the input power.

Then for any vehicle employing stored energy E_s, the system efficiency can be defined as:

$$e = \frac{\Delta KE}{E_m + E_s} \qquad (10.16)$$

For a resonant microwave cavity, the stored energy is given by:

$$E_s = Q_l \text{El} = \frac{Q_l P}{f_o} \qquad (10.17)$$

where Q_l is the loaded Q, El is the electrical loss per cycle and f_o is the resonant frequency.

Note that the stored energy will decrease during the pulse period, as the KE increases and the loaded Q decreases. However, as the following example will show, when acceleration periods are used which are very much less than the cavity time constant, Q_l can be simplified to the static Q value for the cavity.

Then

$$e = \frac{M(V_2 - V_1)}{KPt + \dfrac{2Q_l P}{fo(V_1 + V_2)}} \qquad (10.18)$$

We must now consider what happens in short pulse operation of a high-Q cavity. It results in a reduction of input power due to the rejection of out-of-band components, as shown in Figure 10.15.

Figure 10.15 shows the frequency spectrum for the flight thruster, described in Chapter 7, operating at 3.85 GHz with 1 MHz square wave modulation applied. The pulse length, and therefore the acceleration period, is therefore 0.5 microsecond. Although the main spectral peak is within the 3 dB bandwidth of the cavity, a number of smaller components lie outside the bandwidth, thus reducing the input power to the cavity. The voltage and power ratios of the first to seventh sidebands, designated H1 to H7, are given in Table 10.2.

So although some of the input power is rejected, the stored energy of the cavity builds up to its maximum level due to the time constant of the cavity, designated T. This can be seen in Figure 10.16, which illustrates the theoretical stored energy in the flight thruster cavity for continuous wave (CW) operation, 1 MHz square wave, and 102 kHz square wave modulation. For comparison, the CW stored energy level is set to 0.5. Thus the 102 kHz peak energy level approaches 1.0 over the input pulse, which

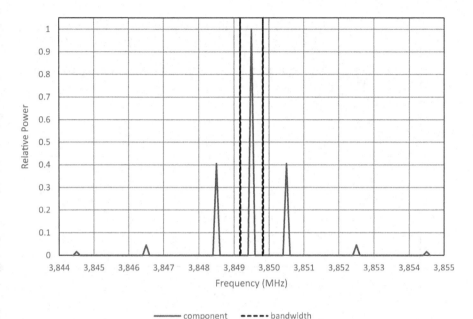

FIGURE 10.15 Spectrum for flight thruster.

TABLE 10.2
Sideband Voltage and Power Ratios

	Fo	H1	H3	H5	H7
Voltage ratio	1	0.637	−0.212	0.127	−0.091
Power ratio	1	0.405	0.045	0.016	0.008

is 5 times the cavity time constant, i.e. 5T. However for 1 MHz modulation, where the input pulse is 0.51T, the peak energy level approaches 1.0 over a longer period.

The theoretical and measured cavity power levels for CW, 102 kHz and 1 MHz modulation are given in Table 10.3.

Thus, the CW stored energy is set to 0.5 and the peak cavity power is 1.0, but for the 1 MHz modulation, the peak power drops to 0.513 due to sideband loss. However, over a time period much greater than the cavity time constant for 1 MHz modulation, the cavity power level rises to a theoretical 1.028 level, which was confirmed by measurement with a 2% error. The practical result is that for 4G short pulse operation, the actual stored energy levels, and thus specific thrust, will be higher than for long pulse 3G operation, as previously illustrated in Figure 10.13.

If we consider equation (10.18), for a constant value of t, the thruster efficiency will increase with the absolute values of initial and final velocities. This is due to the average velocity term $\dfrac{(V_2 + V_1)}{2}$ in the equation (10.15) for input mechanical energy.

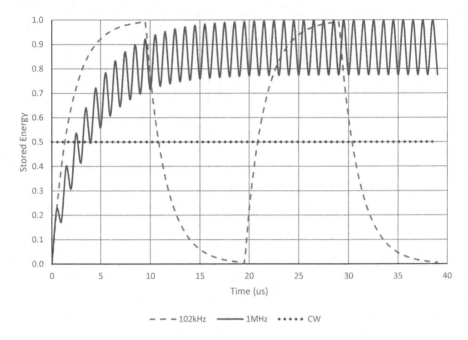

FIGURE 10.16 Stored energy in the flight thruster.

TABLE 10.3
Theoretical and Measured Power Levels in Flight Thruster

	From Theory			Experimental data for nominal 0 dBm input		
Input	Stored Energy	Peak Power Ratio	Q	Theoretical Pdet (mW)	Measured Pdet (mW)	Error, %
CW	0.5	1	23,690	1	1	
102 kHz	0.475	1	21,997	0.882	0.845	−4
1 MHz	0.887	0.513	26,780	1.028	1.01	−2

Thus, the faster the vehicle is moving, the higher the efficiency, and the sooner ΔKE will increase, causing a loss of stored energy and a loss of thrust. Thus, as the velocity of the vehicle increases, the pulse length t must decrease to maintain optimum efficiency. This is illustrated by calculating the pulse length required to maintain 1g acceleration for a proposed heavy launch vehicle, which will be described in detail in Chapter 13.

Figure 10.17 shows that as the vehicle velocity increases from 0.1 km/s to 100 km/s, the pulse length must decrease from 800 microseconds to 0.8 microseconds to maintain an acceleration of 1 g. The thruster efficiency over this range varies from 16.5% to 17.9%.

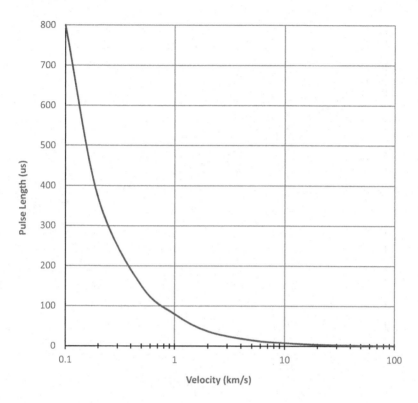

FIGURE 10.17   Pulse length to maintain 1g acceleration.

With the vehicle travelling at 100 km/s, thrust, efficiency and acceleration are plotted against pulse length in Figure 10.18.

Throughout this chapter, we have investigated the effect of acceleration on the very high-Q cavities in superconducting thrusters and have arrived at a method of Doppler compensation and short pulse operation resulting in a fourth-generation thruster design. However, there is also a very important acceleration effect that occurs in the low-Q cavities of first-generation thrusters. But first, it is worth reviewing the reports of further experiments, particularly the anomalies that appeared to have occurred during the first flight tests.

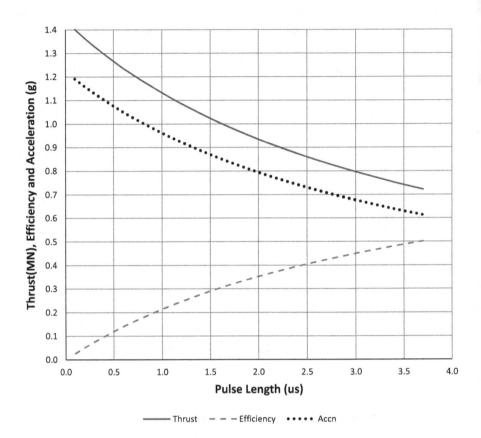

FIGURE 10.18 Heavy launch vehicle performance at 100 km/s.

REFERENCES

1. Shawyer R. The Dynamic Operation of a High Q EmDrive Microwave Thruster. IAC-13,C4,P,44,p1.x17254.
2. Shawyer R. The EmDrive Programme - Implications for the Future of the Aerospace Industry. *CEAS* 2009 Session 11 27 October 2009.
3. Shawyer R. Second Generation EmDrive Propulsion Applied to SSTO Launcher and Interstellar Probe. *Acta Astronautica*.116 (2015) 166–174.
4. Shawyer R. *Superconducting microwave radiation thruster.* UK patent GB2537119.
5. Shawyer R. EmDrive Thrust/load Characteristics. Theory, Experimental Results and a Moon Mission. IAC-19-C4.10.14.
6. Preliminary Assessment. Unidentified Arial Phenomena. Office of the Director of National Intelligence.25 June 2021.
7. Anomalous Threats. Combat Aircraft Journal. April 2021,p16-p21.
8. Schlesinger RJ. *Principles of Electronic Warfare.* Prentice-Hall 1961. ISBN 0-932146-01-5.

11 The First Flights

The very first flight of an EmDrive thruster that has been shown in the public domain can be claimed by the Demonstrator Engine, which on October 31, 2006, at 11:44 am, flew on an air bearing for a total distance of 185 cm [1]. The mean thrust of 9.8 g produced a maximum velocity of 2 cm/s for a 100 kg total mass. One very important parameter measured for each of these flight tests was the friction load of the bearing, which was calibrated before each test. For this test run, the equivalent average load at the centre line of the thruster due to the bearing friction was 8.2 g. The remaining 1.6 g therefore produced the low acceleration that was measured. Although it was not realised at the time, the relatively high ratio of load to thrust ensured that the change in kinetic energy (ΔKE) being extracted from the stored energy of the cavity was kept low. The efficiency of the thruster was therefore low, but the stored energy and, therefore, the thrust level were kept to a maximum. The importance of ensuring that there is a preload force against which the thruster operates was also missed by many research teams trying to reproduce EmDrive operation. This was probably the reason for the anomalies first seen in reports of a Russian flight test of the Yubileiny satellite in April 2009 [2]. It was suggested by defence contacts in the US that this small satellite, propelled by an unknown but novel propulsion system, could actually be the first Russian flight of an EmDrive-type thruster. The 15-year life and 300,000 restarts are typical specification requirements for a TWTA. The satellite mass and power specifications would indicate an X-Band thruster operating at around 40 watts, which is again typical for a military TWTA.

Further reports of possible in-orbit tests using the American X-37B spaceplane, were originally published by The International Business Times on November 7, 2016 [3]. The original tip-off was said to have come from a NASA employee. Again, it was inferred that anomalous results were obtained, although all our US contacts were very tight-lipped about the reports.

These reports were very quickly followed by a press conference held at the Chinese Academy of Space Technology (CAST) on December 13, 2016 [4], claiming that an EmDrive-type thruster had been tested aboard the Tiangong-2 space station. I saw a model of the original Tiangong-1 space station at the IAC conference in Toronto in 2014, and a photo of this model is shown in Figure 11.1.

Although these reports were rightly treated with some scepticism, they each included references to the need for further testing, indicating anomalies were recorded. We were to discover that this was not surprising if some form of preload provision had not been made.

Naturally, this preload requirement is also necessary in any ground-test experimental apparatus. It is therefore ironic that a number of experimenters have concentrated on reducing the loading of their test balances to produce measurement sensitivities down to a few micro-Newtons, in the forlorn hope that this would lead to more reproducible results. NASA first started work on their version of EmDrive at their Eagleworks

FIGURE 11.1   A Model of the Chinese Tiangong Space Station. (Brian Harvey).

laboratory in 2013, although the theory advanced at the time was based on a warp drive concept involving quantum plasma physics. The work was reported in July 2014 [5], where a tapered cavity design was tested on a torsional pendulum. A review of the cavity design, where a number of basic flaws were found, was sent to NASA. However, because the theory of their design did not conform to our theory, the review was ignored. This was unfortunate because the basic NASA design was reproduced by a number of research groups, including Dresden University. Once again, extensive testing failed to produce repeatable results. The Dresden work culminated in a 2022 paper [6]. Because the work had been partly funded by DARPA, we provided a detailed review [7], which showed that there was no possibility of the cavity-producing thrust in any of their tests. This was due to the basic design flaws and the use of the wrong frequencies and modes. A mode analysis was carried out for the 12 test conditions quoted in the paper for the circular cavity, and it was found that only 2 would yield cavity resonance, at 1,819 MHz and 1,959 MHz, but without the dielectric. Clearly, without a difference in guide velocity, which the dielectric would provide, there could be no thrust generated. The apparent resonances that were quoted were attributed to the loop resonating, which explained the low Q values of 2,600 at 1,819 MHz and 2,063 at 1,959 MHz. The geometry of the Dresden tapered cavity, with a flat small end plate and a spherical large end plate, was also analysed. Simple geometry showed that the wavefront phase error reached 180 degrees after 12 transits of the cavity. Now resonance completely disappears when the phase error reaches 180 degrees, implying that the maximum cavity Q value was 6. The Q measurements quoted were simply resonances of the input loop due to a variety of spurious modes. These can always be seen when a high-Q loop is energised in any shape of cavity. They are not measurements of axial resonance between the end plates. The tapered cavity will clearly be incapable of correctly resonating along the axis of the taper, and therefore no thrust will be generated. It is also doubtful that the pendulum-type test apparatus would actually have detected thrust, even if the cavity had been designed and operated correctly.

The group that got closest to reproducing our flight test data was based in Israel and had been granted an export licence by the UK government. Although their cavity was correctly designed and operated, their use of a pendulum balance gave rise to a number of anomalies. The loading problem was not clearly understood at the time, and the work appears to have been abandoned. Our first experience with the loading problem was when we decided to use a pendulum-type test apparatus ourselves. By definition, when the pendulum is in its natural rest position, hanging straight down, there are no sideways forces and therefore no preload on the cavity being suspended from it. A number of test rigs were built, including a sophisticated double suspension pendulum designed and built by an experienced space industry consultancy. Alan and Martin Thompson of Eureco Technologies Ltd. on the Isle of Wight carried out a lengthy programme of careful testing. This produced no discernible thrust from our previously very successful flight thruster.

The flight thruster was then tested on a number of test rigs at the Gilo Industries Group research centre in Dorset. This company, which specialises in powered paragliders and innovative engineering, had the required precision engineering facilities to manufacture a number of different cavity designs. Their adventurous CEO, Gilo Cardozo, had in fact already built a linear air track, and we were anxious to obtain some further test data. Once again, the test results were not compatible with our original test data. However, we had also been given test data from Chinese experiments where an EmDrive thruster was flown on a linear air bearing, with variable results. Their problems were confirmed by our attempts to fly our thruster on the Gilo linear air bearing. At this time, the BBC was filming EmDrive for their Horizon programme on Project Greenglow [8]. For the film sequence, the thruster was simply pushed along the air track.

The flight thruster and a new similar thruster were then tested on a large torsional test rig at Gilo Industries. An extensive, but time-limited, programme was run, but again, the test results were not decisive. Meanwhile, a new superconducting thruster and a large 2.45 GHz thruster were manufactured under the careful supervision of an experienced engineer, Brian Crighton. His background in rotary engines and racing motorbikes made for an easy working relationship, as it reflected my own youthful enthusiasms. At this point, a new government contract was started. However, time and budgetary constraints meant that the work was then moved to another organisation, and the security constraints that had been applied at the very beginning of the EmDrive story were once again restored.

The variability in test data from the NASA, Dresden and Israeli tests, together with the tests at Eureco Technologies and Gilo Industries, had become worrying. Looking back to the reported in-orbit tests, it was realised that there is normally no preload force in the free fall conditions in space and that this was also a common factor in many of the ground test rigs. However, all the successful measurements at SPR had been carried out using test rigs that one way or another, provided a preload or friction load. In an effort to keep costs low, the guiding principle was not to concentrate on the academic approach of very high-resolution force measurements but to accept a typical engineering compromise of allowing a preload while ensuring that it was calibrated.

This finally led to the realisation that there was a fundamental loading theory behind the variable results that needed investigation. The theoretical answer was based on understanding how the stored energy in the cavity will build up following the initial switch-on, under different preload conditions. A much simplified representation of the theory can be given by reverting to the classic bathtub analogy, where three water levels are used as analogies for the stored energy levels resulting from three load conditions. Assume a bathtub with a constant input water flow. Figure 11.2 represents an EmDrive cavity where the preload force is greater than the thrust, and therefore there is no acceleration. The bathtub fills up to level 1, at which point there is enough head of water for the loss through the waste pipe to equal the water input. This is analogous to a stationary cavity where microwave input power equals the losses in the cavity walls and stored energy is at its maximum. The efficiency of the thruster is zero as there is no kinetic energy output.

The second loading condition shown in Figure 11.3 allows the EmDrive cavity to continuously accelerate, enabling the vehicle to build up kinetic energy. In this analogy, the water level builds up to level 2, representing lower stored energy, while continuously filling the second tank, representing the vehicle's kinetic energy.

The third condition, illustrated in Figure 11.4, is where there is zero preload force and the initial acceleration is so high that the instantaneous kinetic energy output equals the input and no stored energy can build up in the cavity. Therefore, no continuous thrust is obtained. In the analogy, the no-load output flow is equal to the input

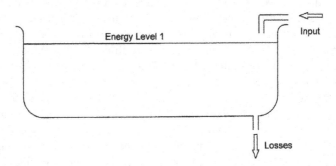

FIGURE 11.2 Bathtub analogy level 1.

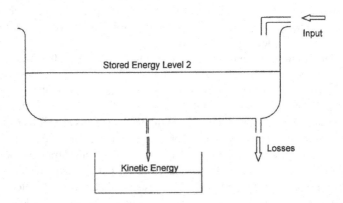

FIGURE 11.3 Bathtub analogy level 2.

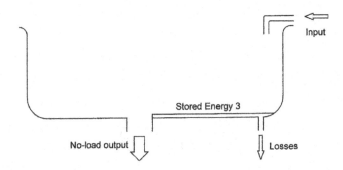

Input

Stored Energy 3

No-load output

Losses

FIGURE 11.4 Bathtub analogy level 3.

flow, water level 3 is negligible, and the bathtub is unable to fill. Therefore, there is negligible stored energy, and no thrust will be measured.

The practical effect of this theory can be seen from the thrust output calculated for a 20 kg satellite operating with a theoretical 1 microgram preload. For this calculation, covering the first 25 microseconds after switch-on, the effect of the inertial load of the satellite is ignored. Figure 11.5 shows the kinetic and stored energy levels and the resultant thrust for this satellite design, which is based on commercially available cubesat components and a 3.85 GHz EmDrive thruster operating at 40 watts power.

Figure 11.5 shows that the stored energy increases according to an exponential function, while the kinetic energy increases according to a square function. At 20 microseconds after switching on, the kinetic energy is increasing faster than the stored energy; therefore, the stored energy can no longer build up and the initial thrust pulse cannot be sustained. The first conclusion to be drawn is that if the cubesat thruster is tested on a test rig that eliminates its inertial load and the preload is less than 1 microgram, thrust will not be continuously measured. The test apparatus must therefore be designed to provide the correct preload force to successfully measure thrust. The second conclusion is that when in orbit, when no preload force is present, loading techniques are required on start-up. Inertial loading will then occur once sufficient acceleration is established. A contact working with a secretive international commercial organisation aiming to conduct in-orbit tests with EmDrive thrusters remarked, *"You need to give it a small push to get it moving"*.

To investigate these interesting results and the theory behind their cause, the flight thruster was returned to SPR Ltd., and a series of thrust measurements were performed against varying loads. The test apparatus is illustrated in Figure 11.6, and the components are identified in the diagram shown in Figure 11.7.

The balance could be used for measuring both thrust and reaction force by carefully offsetting the level of the beam from the horizontal, by varying the stop height. If the offset was kept very small, the calibration factor variation, which was checked before each test, was kept to an acceptable level, as shown in Figure 11.8.

With the thruster mounted in the horizontal balance setting, the balance measured the reaction force, which is an upward force causing the balance reading to decrease. The above and below horizontal settings measured thrust, which is a downward force, causing the balance reading to increase. By setting the balance very near horizontal,

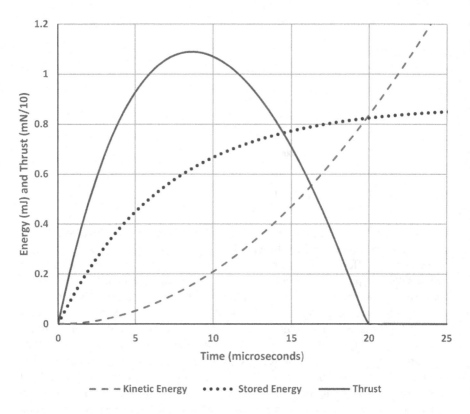

FIGURE 11.5   Cubesat thruster switch-on.

FIGURE 11.6   Flight thruster on preload test rig.

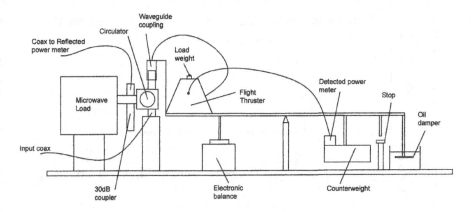

FIGURE 11.7   Diagram of the preload test rig.

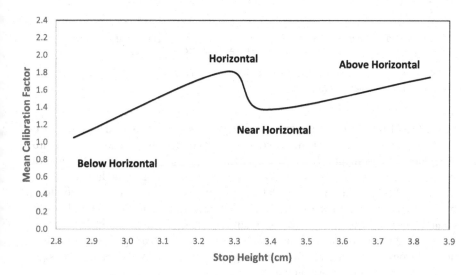

FIGURE 11.8   Beam balance calibration.

both thrust and reaction force could be measured during the same test run. To determine the balance noise due to spurious forces caused by thermal and electromagnetic effects as well as any background vibrations, a routine test was carried out where the thruster was tested off resonance at typical input power levels. Thus, although test conditions were as close to normal as possible, there was no contribution from either thrust or reaction force. Figure 11.9 shows a typical result. With a mean load force of 174 mg, a noise standard deviation of 14 mg was measured, yielding a 12 sigma result.

A series of 109 tests were carried out in this second flight thruster test programme. The tests included 26 high-power, calibrated test runs, which were carried out to measure both thrust and reaction force, with varying preload weights added

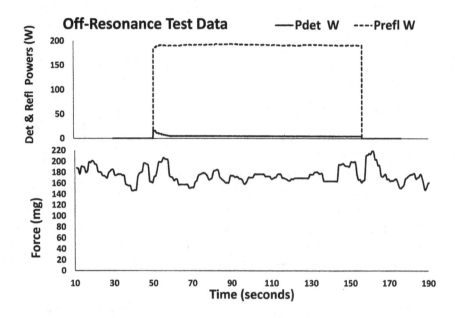

FIGURE 11.9   Noise calibration results.

to the top of the thruster. The main object of these tests was to determine thrust/ load and reaction force/load characteristics for continuous wave (CW) and amplitude modulation (AM) inputs. To allow easy comparison of thrust or reaction force data with applied loads, the values are given in mg. After a series of modulation tests, an AM of a 300 Hz square wave was determined to be the optimum modulation for the mechanical response of the balance. It is interesting to note that the early experimental thruster, originally tested on a similar balance and reported in Chapter 3, was powered by a commercial magnetron. This source had a high-voltage DC input, derived from a half-wave rectifier circuit, giving an approximation to a 50 Hz square wave AM on the microwave output.

The beam balance enabled the load applied to the thruster to be varied by using different balance weights on top of the thruster. The input power to the thruster was maintained close to a constant value for each set of test runs to enable direct force measurements to be compared.

Figure 11.10 shows the results from a set of six test runs with a mean CW power of 167 watts. The balance was set to be marginally below horizontal to enable thrust to be measured. Thrust is in a downward direction, therefore increasing the balance reading and being recorded as a positive value. The results clearly illustrate the theoretical prediction that thrust will approach zero as the applied load approaches zero. Also, as the applied load goes above the maximum thrust measured, the thrust again approaches zero, as predicted by the theory. Maximum thrust is 467 mg at an applied load of 203 mg.

Figure 11.11 shows the results from a set of four test runs with a mean CW power of 162 watts. In this set, the balance was set to exactly horizontal to enable

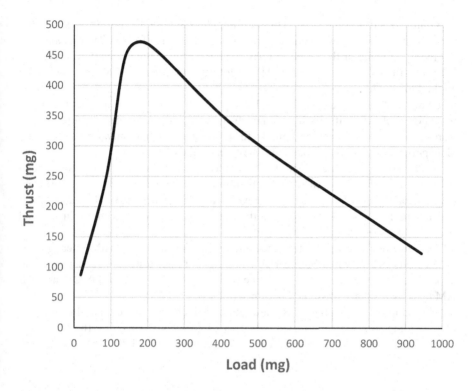

FIGURE 11.10 Thrust/Load results for the mean CW power of 167 watts.

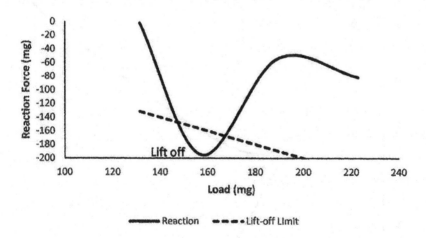

FIGURE 11.11 Reaction force/Load results for the mean CW power of 162 watts.

reaction force to be measured. Clearly, the reaction force is now upwards and is therefore measured as a reduction in balance readings and recorded as a negative value. Once again, as the load approaches zero, the reaction force goes to zero, whereas with increasing load, the reaction force is below the maximum value. The maximum reaction force is −195 mg at an applied load of 157 mg. In this run, the beam lifted off the balance pan as it was below the lift-off limit. This means that the recorded value of the reaction force is necessarily below the actual force. The lift-off limit is shown in Figure 11.11 by plotting reaction force values equal to the applied load values.

Figure 11.12 shows a second set of five reaction force measurements at a lower input power of 125 watts. In this set, no lift-off occurred, as the reaction force values were all below the lift-off limit, and so the data set has no anomaly. Once again, the plot approaches zero as the load approaches each end of the range.

Figure 11.13 shows the effect of AM on the thrust/load characteristic for a mean input power of 155 watts. The 7 test runs show an approximate level value of 300 mg, over the load range of 51 mg to 944 mg with two peak values above 500 mg at 83 mg and 164 mg loads. These peaks are in the same load range of maximum thrusts and reaction forces shown in the CW results.

The reaction force results from the final set of 4 AM test runs, with a mean power of 182 watts, are shown in Figure 11.14. The higher power tests produced a lift-off at a load of 29 mg, with zero reaction force being approached at a load of 280 mg. All AM tests used square wave modulation at 300 Hz.

This series of tests, although somewhat hurried, clearly showed that the concepts of preloading and AM would need to be taken into account when in-orbit tests were carried out. The results were reported in a lecture given at the UK Defence Academy

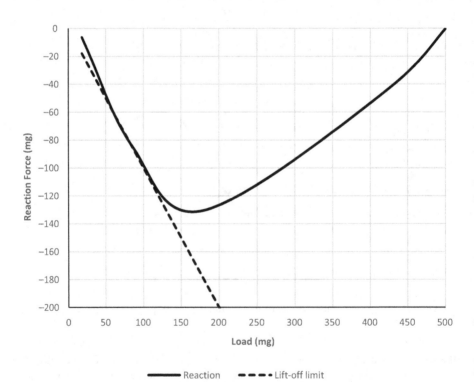

FIGURE 11.12   Reaction force/Load results for the mean CW power of 125 watts.

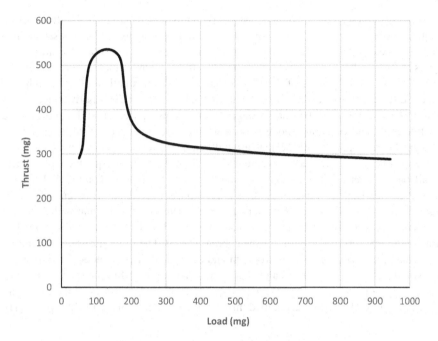

FIGURE 11.13   Thrust/Load results for the mean AM power of 155 watts.

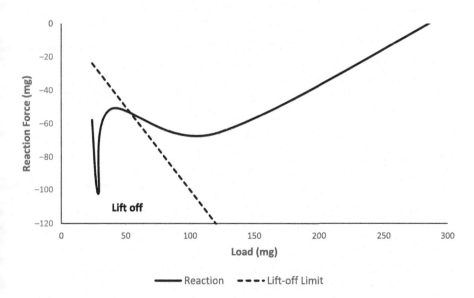

FIGURE 11.14   Reaction force/Load results for the mean AM power of 182 watts.

at Shrivenham on February 12, 2019 [9], and then at the 2019 IAC conference in Washington in October of the same year [10]. While attending the conference, I followed up on an invitation to talk to a number of researchers at DARPA headquarters as well as attend a meeting with the director.

With in-orbit testing of EmDrive on small commercial satellites being discussed in a number of quarters, a thruster design study was carried out for a 12U cubesat application. A cubesat is basically a small satellite whose overall dimensions are constrained by units of 10cm cubes. Thus, a 12U cubesat will nominally have dimensions of 30 cm × 20 cm × 20 cm. The advantage of these dimensional constraints is that a large number of satellite functions are readily available at relatively low cost. The disadvantage is that the EmDrive cavity must be small enough to fit inside the spacecraft body, so the choice of frequency and operating mode is critical. EmDrive propulsion systems have been designed to operate from 900 MHz to around 8 GHz. However, to maintain the low-cost, cubesat philosophy, a frequency where both components and test equipment are widely available is preferred. The first choice would seem to be the ISM band (2.4–2.5 GHz). However, this leads to cavity sizes that are too big for the largest 12U cubesat structure. The original SPR Flight Thruster was designed around a flight-qualified, 3.85 GHz TWTA, and as equipment in the 5G mobile band (3.4–3.8 GHz) is now readily available, this frequency was also chosen for the cubesat design. The flight thruster had a maximum internal cavity diameter of 200 mm and operated in TE013 mode. To reduce the maximum diameter to 150 mm and enable the cavity to fit within the spacecraft body, the operating mode was changed to TE113. The internal dimensions of the cavity are given in Figure 11.15, and the detailed thruster design is given in a 2020 IAC paper [11].

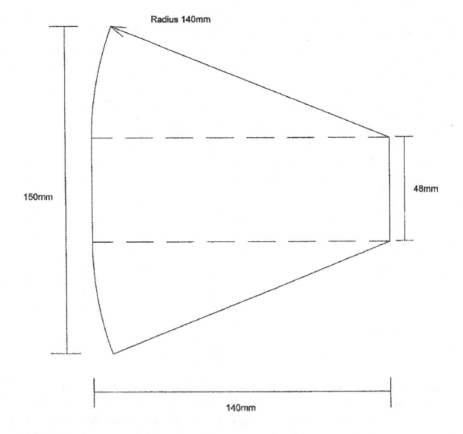

FIGURE 11.15   Internal dimensions of the cubesat cavity.

The low power available on a typical cubesat means that the thruster design must utilise the power in the most efficient manner. Essentially, this requires that the cavity Q must be maximised. This requirement is addressed in two ways. Firstly, the cavity is cooled by ensuring that the large end plate of the cavity is directed into deep space and is kept in the shade using a deployable sunshade. Further thermal cooling panels are also deployed. Secondly, the end plate alignment, which is critical for high Q and high thrust, is kept within tolerance by the use of the SPR Flight Thruster cavity geometry. This enables the critical dimensions to be mainly determined by flat surfaces, which are easier to machine and measure. A Q value of 84,000 and a specific thrust of 0.5 N/kW were specified for the thruster. Then, for 40 watts input power, the thrust is 20 mN. The operating temperature of the cavity under optimum attitude conditions was predicted to be –75°C. The fully deployed configuration of the satellite is shown in Figure 11.16.

Figure 11.16 shows that the satellite carries thermal radiating panels, A, B, C, D and E, to reduce the operating temperature of the thruster, which will dissipate a maximum of 40 watts. The A panel is shaded using a lightweight deployable sunshield, while the other panels are shaded by the solar arrays. The SSPA is thermally isolated from the thruster and is cooled by panel F to a maximum operating temperature of 61°C while dissipating 52 watts. Panel F is shaded by the main body of the satellite.

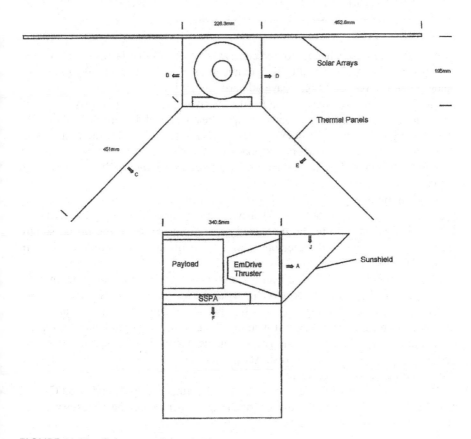

FIGURE 11.16   Cubesat configuration.

TABLE 11.1
Cubesat Mass Budget

Item	Mass (kg)
Cubesat structure	1.386
Solar panel	0.932
Array structure	2.311
Power conditioner	0.148
Battery	0.67
Frequency generator	0.019
SSPA	0.856
Attitude control	1.135
Telemetry & command	0.1
EmDrive cavity	2.4
EmDrive control unit	0.3
Cabling	0.2
Thermal structure	2.233
10% contingency	1.269
Payload	6
Total	19.959

A preliminary mass budget for the cubesat is given in Table 11.1, where a margin of only 10% is added as most of the equipment is already flight-qualified. To keep below a launch mass of 20 kg, a payload mass of 6 kg is allocated.

In-orbit operation of the cubesat would clearly require the previously described problem of providing a preload on start-up to be addressed. Figure 11.17 shows the stored energy (E_s) and kinetic energy (E_k) levels in the cavity for the first 35 seconds after switch-on for varying preloads (W). This period represents five times the time constant of the cavity, during which Es rises exponentially to a maximum value.

The kinetic energy of the total inertial load of 20 kg is shown for increasing pre-loads from 2 μg to 500 μg. Clearly, at a pre-load of 2 μg, the increase in kinetic energy is close to the increase in stored energy once the initial square law rise is modified by the inertial load. The resulting thrust profile is given in Figure 11.18.

The thrust response for a pre-load of 2 μg gives a small peak during the initial 5 microseconds when stored energy is slightly above kinetic energy. Thrust then drops to zero as kinetic energy closely follows the rise in stored energy, thus leaving the net energy available for generating thrust approaching zero. As the pre-load is increased, the initial acceleration drops, the rise in kinetic energy is reduced, and more net energy is available to produce thrust. At a pre-load above 500 μg, i.e., a force of 5 μN, constant thrust will be available.

Clearly, a cubesat with only one thruster and a simple Attitude and Orbit Control System will need to adopt additional techniques to provide the necessary initial

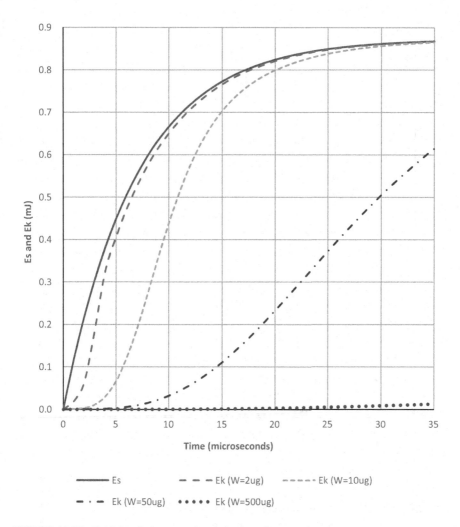

FIGURE 11.17 Initial switch energy levels in the cubesat cavity.

preload. Although the flight model tests indicated that AM offers a solution, to optimise the power system and SSPA mass, the SSPA and thruster should be run in
CW mode, and so mechanical methods of applying the preload were investigated. A
momentum wheel can be used to induce a pitch or yaw movement, which will produce a centrifugal force on the cavity. For this to be effective, the thruster must be
positioned along the Z axis with sufficient separation from the spacecraft's centre of
mass. An initial mass properties analysis gives this radius as 146 mm, which means
the required pre-load of 5 μN can be obtained with an angular velocity of 1.3 mrad/s.

The use of gravity loading while in LEO can also be applied. In this case, the
cubesat would be oriented with the z axis aligned to the Earth's gravity vector, with
the cavity mass below the centre of mass of the cubesat. The pre-load force obtained

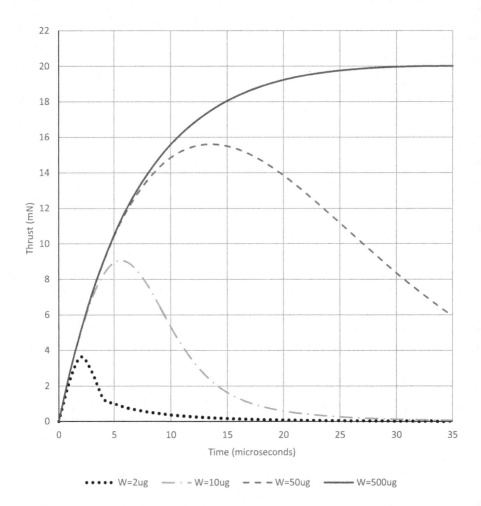

FIGURE 11.18 Initial thrust profile for a cubesat thruster.

from the gravity gradient within the cubesat is shown in Figure 11.19. Up to an orbital altitude of 1,800 km, the gravity gradient will produce a continuous pre-load force above the required 5 μN.

Alternatively, to avoid the use of the Attitude and Orbit Control System system, a small solenoid with the body fixed to the spacecraft structure and the armature free to move can be used. The solenoid stroke is aligned along the thrust axis. The solenoid is pulsed as the thruster is switched on, and the armature momentum is mechanically transferred to the thruster through the spacecraft structure. This mechanical impulse provides a pre-load force pulse while the armature is moving, and the Q value and thrust will continue to build up during this period. Once this start-up impulse is complete, momentum is reversed, but the inertial load of the accelerating spacecraft ensures thrust is maintained. The solenoid can then be switched off, returning the armature to its original position. A typical solenoid can transfer an average pre-load

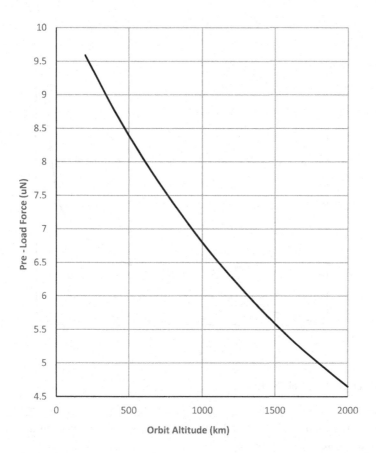

FIGURE 11.19 Gravity gradient preload.

force of 5 μN to the spacecraft over a period of 10 milliseconds for an input power of
2 watts. A solenoid pre-load force profile is shown in Figure 11.20.

The solenoid would be incorporated in the EmDrive control unit, which also
includes a frequency generator circuit and a processor to control the thruster input
frequency from the feedback of power and temperature monitoring circuits. Start-up
frequency sequencing to optimise frequency for the differing temperature profiles of
the cavity and input circuit is essential and is controlled by the processor. A sweep
function would also be included to enable Q measurements to be made as part of routine performance monitoring. Thruster telemetry and command functions are routed
through the EmDrive control unit.

Although the primary purpose of the design study was to provide a low-cost demonstration of first-generation EmDrive operating in LEO, the possibilities of more
ambitious missions using the cubesat were too enticing not to consider. The design
results in a 20 kg spacecraft propelled by a thruster that provides 20 mN of continuous thrust. Assuming constant illumination of the solar arrays, this level of thrust
could be maintained over a typical cubesat-specified operating lifetime of 5 years. A

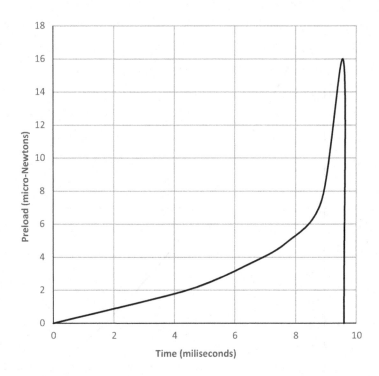

FIGURE 11.20   Solenoid preload force.

maximum terminal velocity of 160 km/s is therefore possible, putting modest deep space science missions within reach.

However, the first mission application to be considered was the commercial use of the cubesat as a space tug for small satellites. If the payload is a 50 kg satellite, a total delta V of 46 km/s can be achieved. This can be compared to a delta V of only 1 km/s for the same payload, which is the limit of the existing best small electric space tug performance using microwave power and water propellant. Clearly, with this huge improvement in performance and a suitable docking mechanism, the EmDrive-propelled cubesat could carry out multiple space tug missions in Earth orbit or lunar transfer missions. For the transfer from LEO to low Lunar orbit of a 100 kg satellite, an outward flight time of 18 months is required. With a return flight time of 3 months, three transfer missions are possible within the 5-year operational life, assuming the 20 kg cubesat is left in low Lunar orbit on the third mission. This can again be compared with a microwave and water-propelled spacecraft, which would weigh 450 kg and only manage a single transfer, taking 10 months.

Two interplanetary missions illustrate the flight times possible for an EmDrive-propelled cubesat. The first is a flight from LEO to a low Mars orbit. The flight starts with a spiral out of LEO to a Mars transfer orbit, taking 64 days. A direct transit is followed by acceleration to the midpoint and then deceleration to Mars capture. This takes 152 days, during which the spacecraft reaches a maximum velocity of 11 km/s. Mars capture takes a further 10 days and is followed by a spiral down to low

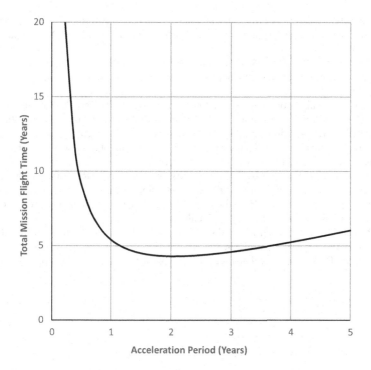

FIGURE 11.21   Pluto fly-by mission.

Mars orbit of 21 days. The estimated total mission time is based on a total delta V requirement of 6.6 km/s and mean eclipse periods of 0.32%, resulting in a flight time of 8.1 months.

However, the true potential of EmDrive interplanetary missions is illustrated by the possibility of a Pluto flyby mission within a 5-year total flight time. Such missions need to maintain acceleration periods within 1 AU distance from the sun to provide nominal solar array performance. The mission profile thus becomes a slow spiral out of LEO, an acceleration period in multiple Sun orbits, plus a cruise period at a constant terminal velocity.

Figure 11.21 gives the result of a simple mission optimisation to give a minimum total mission time of 4.3 years for an acceleration period of 2 years and a terminal velocity of 63 km/s. However, to travel any further into space and still maintain reasonable project timescales, the use of a superconducting thruster becomes mandatory.

REFERENCES

1. Emdrive - Home. June 2015. Dynamic Test. Full DM test 188. www.emdrive.com.
2. Pravda articles. https://english.pravda.ru/science/tech/14-04-2009/107399-Russian_scientists-0 https://english.pravda.ru/science/tech/18-02-2010/112279-nano_satellites-0

3. International Business Times article. Mary-Ann Russon November 2016 EmDrive: US and China already testing microwave thruster on Tiangong-2 and X37B space plane (ibtimes.co.uk)

4. International Business Times article. Mary-Ann Russon EmDrive: Chinese space agency to put controversial microwave thruster onto satellites 'as soon as possible' (ibtimes.co.uk)

5. Brady DA et al. Anomalous Thrust Production from an RF Test Device Measured on a Low-Thrust Torsion Pendulum. NASA Lyndon B. Johnson Space Center. 50th AIAA/ASME/SAE/ASEE Joint Propulsion Conference July 28-30, 2014

6. Neunziga O, Weikertb M, Tajmar M. Thrust Measurements of Microwave- superconducting and Laser-Type Emdrives. IAC-21,C4,10-C3.5,1,x63502

7. Analysis of TU Dresden IAC-21 test results. Note sent to DARPA 10 January 2022.

8. BBC Horizon TV documentary. Project Greenglow.2016.

9. Shawyer R. The impact of EmDrive on Future Warfare. Edited copy of presentation given at UK Defence Academy Shrivenham 12th February 2019.

10. Shawyer R. EmDrive Thrust/load Characteristics. Theory, Experimental Results and a Moon Mission. IAC-19-C4.10.14.

11. Shawyer R. An EmDrive Thruster for Cubesats. IAC-20-C4.2.6. 56845.

12 Deep Space Missions

Having hopefully made the case for EmDrive's viability, we will now look at three future vehicles based on a common fourth-generation engine. The work leading up to this next-generation technology is reported in Chapter 10 and includes the development of a method of Doppler compensation and short pulse operation. The first of the new vehicle designs is a deep space probe using a single engine.

Deep space science missions, where the vehicle is sent over huge distances with only data being returned, clearly need high terminal velocities. The vehicle specification must aim for simplicity and low cost, in addition to high speed. The last requirement is obviously to enable the flight times to be minimised so that, for instance, science missions can be accomplished within the working lifetime of the principle scientists involved. Note that for these missions, simply defining and agreeing on the objectives can take considerable time, let alone the payload equipment development schedules. The process could be simplified by having a commercially available launch vehicle developed for future industrial operations such as asteroid mining. This will require many fast prospecting missions before committing to the actual mining itself. A particular challenge with superconducting EmDrive propulsion is clearly the need to carry liquid hydrogen, which has the annoying characteristic of leakage and boil-off. This implies short propulsion periods measured in hours or days and therefore high rates of acceleration. The vehicle proposed for such applications is the deep space probe illustrated in Figure 12.1.

The probe uses one of the 950 MHz engines proposed for the heavy launch vehicle and the personal spaceplane, to be described in Chapters 13 and 14. This engine design has evolved from that used in the Hybrid Spaceplane, illustrated in Figure 10.6, where, although a very high specific thrust of 9.92 kN/kW was predicted, the acceleration was limited to 0.5 m/s/s. The Doppler shift model, described in Chapter 10, led to a significant change in the mode used in the 950 MHz engine compared to the flight model thruster. As well as using a TE21 mode to reduce cavity diameters, the axial length was reduced by limiting p to 1. Thus, the cavity resonates with only one half wavelength. The longer axial lengths used in the experimental thrusters, where $p = 3$, were adopted to minimise the E and H field distortions caused by the input loop. With careful design, the input helix antenna used in the superconducting cavity, illustrated in Figure 10.9, causes less distortion and enables a $p = 1$ design to be realised. The importance of the axial length reduction is illustrated by equation (10.1),

where the transit time, $t_s = \dfrac{p}{2Fo}$.

The Doppler model equations (10.1) through (10.7) show that the increase in transit time leads to an increase in Doppler Shift and therefore higher input frequency and axial length compensation requirements.

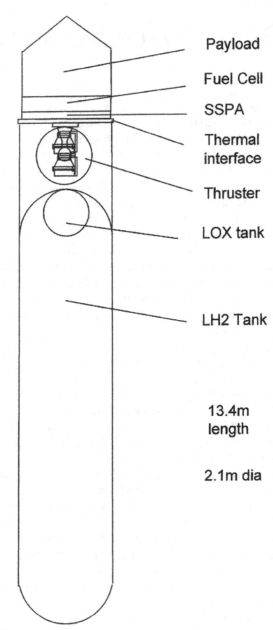

FIGURE 12.1   Deep space probe.

   The TE211 mode and a nominal frequency of 950 MHz resulted in internal diameters
of 500 mm for the large end plate and 308 mm for the small end plate. The axial length
for resonance was 261.9 mm. The E and H field phase plots are given in Figure 12.2.

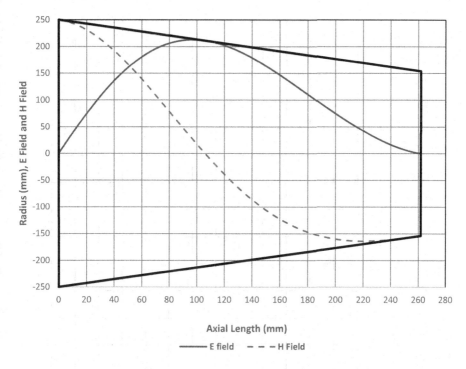

FIGURE 12.2 E field and H field phase responses for a 950 MHz cavity.

The single half-wave response of the fields can be compared to the 3 half-wave response for the flight cavity, shown in Figures 7.2 and 7.3. The guide wavelength and wave impedance plots are given in Figure 12.3 and may be compared with the plots in Figure 7.4 for the flight cavity.

The predicted Q value at 20°K is 7.98×10^8, and together with a design factor of 0.7511, equation (9.2) gives a specific thrust of 4 kN/kW.

The use of a single engine limits the nominal launch mass to 5 tonnes for the deep space probe. The probe can be launched in space from an uncrewed spaceplane, or up to 10 probes can be launched from a heavy launch vehicle. Of particular interest for low-cost science missions or for large numbers of asteroid prospecting payloads will be the ability to launch the vehicle from a launch tube mounted on the back of a truck. To ease launch operations, the tube can consist of a double-walled vacuum tube with the probe inserted up to the thermal interface plate. The arrangement mini-mises boil-off from the LH2 tank while allowing access to the ambient temperature equipment and payload during final pre-launch checks. When elevated to the verti-cal position on the truck, the vehicle mass is supported by the thermal interface plate resting on the launch tube top flange. This minimises tank mass as the com-bined LH2 and LOX tank simply hangs from the interface plate and is therefore not required to support the total vehicle mass. With the vehicle's centre of mass below the fully gimballed engine, attitude control of the vehicle is readily achieved under weather conditions that would normally prohibit launch. Roll control is carried out

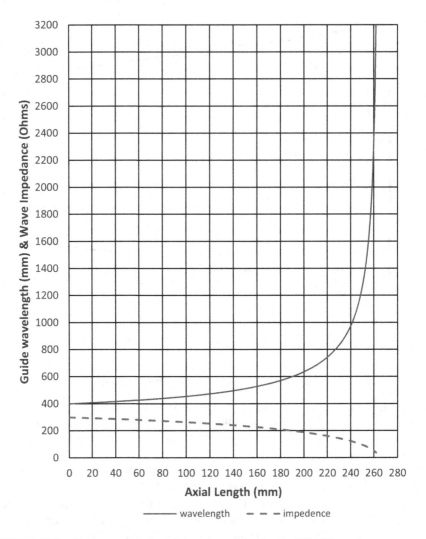

FIGURE 12.3 Guide wavelength and wave impedance for the 950 MHz cavity.

using a momentum wheel during all phases of flight. The ground launch configuration is shown in Figure 12.4.

An outline of the mass budget is given in Table 12.1. The engine bay mass includes thrusters, gimbals, enclosures and thermal interface mounting.

Mission analyses have been carried out for an in-space launch based on a thruster input power of 80 kW, resulting in a thrust of 320 kN. The analysis technique uses a simple spread sheet approach with time increment and vehicle acceleration as inputs in the first two columns. Mission parameters are then calculated for increasing flight time using standard equations as listed in Table 12.2. LH2 usage is calculated from microwave energy increments, using the latent heat of evaporation of LH2 to maintain 20°K. The LOX usage is based on the DC power requirements of the SSPA.

FIGURE 12.4   DSP ground launch configuration (Jung Dong Whi).

TABLE 12.1
Deep Space Probe Mass Budget

Item	Mass (kg)
Payload	500
Engine bay	795
SSPA	758
Fuel cell	524
Fuel	2,500
Tanks	310
Total	5,387

For Thrust $= T$ kN
Microwave Power $= P$ kWm
Launch Mass $= M_L$ kg
Launch Fuel Mass $= M_F$ kg
Initial Velocity $= V_i$ m/s
Latent heat of evaporation of LH2 $= H$ kJ/kg
LOX flow $= F_o$ kg/kWh.
SSPA efficiency $= e_a$

Mission analysis equations are given in Table 12.2.

Note that to avoid a circular calculation, for an acceleration value at Time Increment (n), the value of M_F must be used for Time Increment ($n-1$).

For an in-space launch with a payload mass of 500 kg, the distance and terminal velocity over a full propulsion period of 185 minutes are given in Figure 12.5. Note that payload mass can be increased by partial filling of the fuel tanks and trading off the fuel margin (F_m) shown in Figure 12.6 for additional payload mass. Thus, for

TABLE 12.2
Mission Analysis Equations for In-space Launch

Column	Parameter	Symbol	Unit	Equation
A	Time increment	t	s	Input
B	Acceleration	a	m/s/s	$a = 1000T/(M_L - M_F)$
C	Terminal velocity	V_t	m/s	$V_t = V_i + ta/1000$
D	Velocity Increment	ΔV	m/s	$\Delta V = V_t - V_i$
E	Average velocity	V_{av}	m/s	$V_{av} = (V_t + V_i)/2$
F	Total distance	d	km	$d = d_i + tVav$
G	Microwave energy increment	ΔE_m	kJ	$\Delta E_m = Pt$
H	LH2 mass increment	ΔM_H	kg	$\Delta M_H = \Delta E_m/H$
I	DC energy increment	ΔE_{dc}	kWh	$\Delta E_{dc} = \Delta E_m/3600\, e_a$
J	LOX mass increment	ΔM_o	kg	$\Delta M_o = F_o\, \Delta E_{dc}$
K	Total fuel used	M_F	kg	$m_F = \Sigma(\Delta M_H + \Delta M_o)$
L	Fuel margin	F_m	kg	$F_m = M_F - m_F$
M	Total propulsion period	T_P	Minutes	$T_P = \Sigma(t/60)$

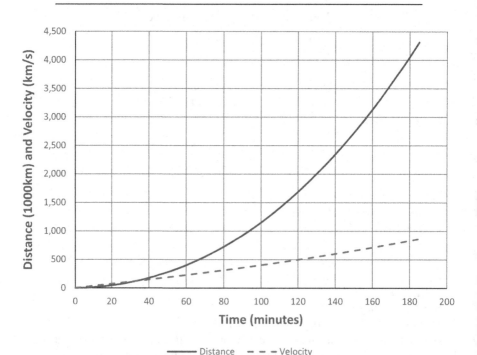

FIGURE 12.5 Space launched deep space probe.

example, Figures 12.5 and 12.6 show that an additional 1000 kg of payload can be gained by limiting the propulsion period to 110 minutes and reducing the terminal velocity to 450 km/s.

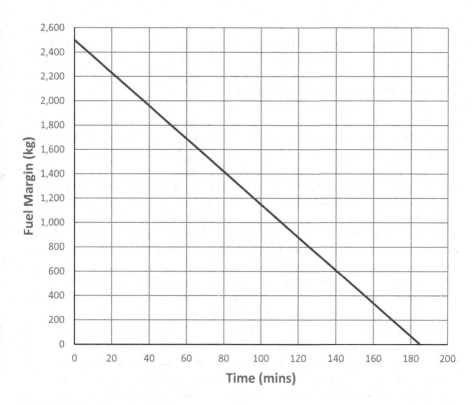

FIGURE 12.6   Fuel margin for in-space launch.

For the nominal payload mass of 500 kg, Figure 12.5 shows a terminal velocity of 863 km/s is achieved after 185 minutes of thrust. The initial acceleration is 6.1 g, and with the full fuel load of 2,500 kg used up, the final acceleration is 11.3 g. The distance covered during the thrust period was 4.3 million km. Thus, for a final vehicle mass of 2,887 kg, the kinetic energy of the vehicle will be 1.08×10^9 MJ. For a constant thrust of 320 kN, the mechanical energy input over the propulsion distance is 1.38×10^9 MJ, the total dc electrical energy input is 1.97×10^3 MJ, and the total thermal cooling energy from 1,952 kg of LH2 is 8.88×10^2 MJ. Thus, the overall thruster efficiency for the mission is 0.783. This can be compared to the orbital thruster efficiency of 0.243 calculated for the SSTO spaceplane described in Chapter 10, with spaceplane parameters given in Table 10.1. Clearly, the Doppler compensation and short pulse operation of the fourth-generation thruster not only allow a considerable increase in acceleration but also improve the overall mission efficiency.

Once the thrust period is finished, a cruise phase of 402 days will take the probe out to a distance of 200 AU, which is generally regarded as the outer limit of the heliosphere, called the heliopause, and the beginning of interstellar space. The heliosphere is the vast volume of fields and particles streaming out from the sun on the solar wind. In spite of its name, the heliosphere is not a sphere but a comet-like, elongated shape with a nose and a tail. This flight time of just over a year can be compared with the shortest time of 23.8 years predicted for a conventional electric propulsion probe [1]. The paper proposed solar electric propulsion for the first stage and a radioisotope thermal generator-powered second stage, launched by an Arianne 5 rocket. The minimum flight time also required a gravity assist fly-by of Jupiter, which reduced the launch window to once every 10 years if the escape direction was restricted to the heliopause "nose". This long mission time does not present an attractive project for even a young scientist to join.

Typical deep space science missions, with overall costs within modest government or university science budgets, can be envisioned by using a ground-launched probe. In this case, the terminal velocity for a fly-by mission is reduced as fuel is used to overcome drag during the initial climb through the atmosphere, and thrust is also required to overcome the Earth's gravity well as the probe climbs away from the Earth. These effects must be included in additional mission analysis equations. The drop in vehicle acceleration is shown in Figure 12.7.

It is interesting to note the difference in thruster efficiency for both ground and space launch of the DSP. This is shown in Figure 12.8, where the drag and Earth gravity effects cause a reduction in the transfer from cavity stored energy to kinetic energy and thus a reduction in efficiency. However, the final velocity for a ground-launched 500 kg payload is only marginally reduced to 856 km/s, compared to 863 km/s for a space launch. The total distance travelled during the propulsion period is also marginally reduced to 4.2 million km from 4.3 million km.

An ideal application for ground-launched probes would be prospecting asteroids for potential mining operations. There have been many proposals suggesting that huge fortunes can be made by mining the precious metals that asteroids are said to contain. Once the right asteroid is found, what is mining likely to yield? It has been estimated that a 45-m-wide asteroid weighing around 1 million tonnes could contain

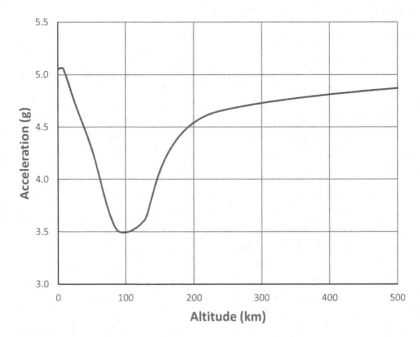

FIGURE 12.7 Initial climb of the ground-launched probe.

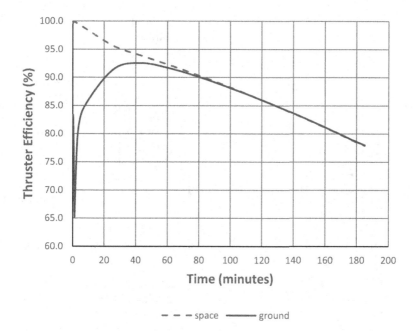

FIGURE 12.8 Thruster efficiencies for space and ground-launched DSP.

30 tonnes of precious metals [2]. These are principally the six platinum group metals: rhodium, palladium, iridium, platinum, osmium and ruthenium. They are listed in descending value and together represent a current value of $36 million per tonne. Therefore, the total value of completely mining and extracting these metals from just one relatively small asteroid is $1 billion. EmDrive heavy launch vehicles with a 50-tonne payload capacity would be used for both transporting the mining and extraction equipment to the asteroid and carrying the metals back to Earth. Clearly, finding the right type of asteroid is worth sending out a large number of probes to survey its likely prospects.

This means the probe payload must first be placed in orbit around the asteroid. The core region of the main asteroid belt lies between 2.06 AU and 3.27 AU and contains 93% of all asteroids. A typical asteroid prospecting mission to the middle of this belt would therefore need to travel 1.7 AU from the Earth, i.e., 255 million km. Assuming a ground launch and a 500 kg payload, the probe would accelerate for 1.6 hours to reach a velocity of 363 km/s after a distance of 0.96 million km. The thruster would then be turned off for a cruise period of 194 hours before a deceleration period of 1.3 hours brought the terminal velocity to zero, enabling the probe to go into orbit around the asteroid. A fuel margin of 215 kg remains for LH2 boil-off and manoeuvring in orbit. The total mission time of just over 8 days from a simple ground launch carrying a 500 kg payload is going to be a very attractive proposition for the large number of asteroid prospecting missions necessary to start the asteroid mining industry. Once an orbital inspection of the asteroid determines that a lander is worth sending, then a modified Deep Space Probe (DSP) with a lander payload, as shown in Figure 12.9, would be launched.

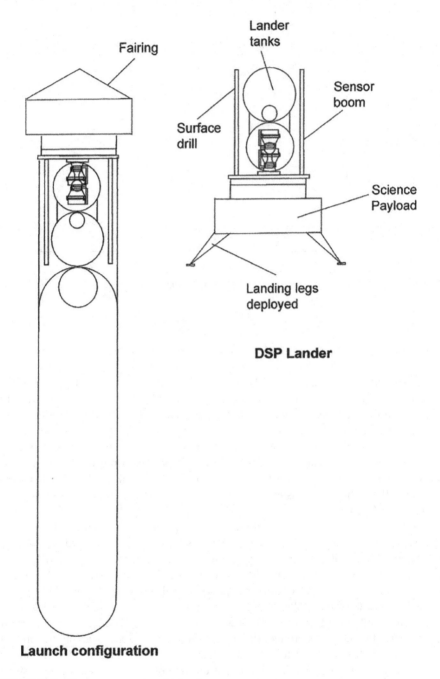

FIGURE 12.9 Deep space probe with lander.

With the thruster rotated 180 degrees from the launch attitude, the DSP is decelerated to a low approach velocity. The Lander would then detach from the main body of the DSP at the thermal interface, and the thruster would provide the remaining deceleration to the lander only. Note that the centre of mass of the lander alone is above the thruster in the launch configuration, whereas the centre of mass of the whole vehicle was below the thruster at launch. In the landing attitude the centre of mass of the lander is below the thruster, thus producing the same inherently stable configuration as during the launch. With the fairing detached and the landing legs deployed, the thruster would then gently lower the lander to the surface of the asteroid. There is space within the DSP body for drill sections and sensor booms up to 2.7 m long, to be carried on the lander. These would be deployed on the asteroid surface to allow a detailed geophysical survey to be carried out at the landing position. The science payload would analyse the drill samples and communicate data on the precious metal content to Earth. The drill and sensor booms can then be stowed, and the lander can move to other places on the surface to complete a full survey of the asteroid.

Finally, we cannot leave the topic of the deep space probe without considering the existential question of how to deflect any incoming asteroid on course to hit the Earth. Major asteroid strikes, called meteorites when they actually reach the Earth, have been recorded throughout the history of the planet. The latest, in February 2013 at Chelyabinsk, estimated to be around 20-m wide and travelling at 19 km/s, caused an explosion equivalent to half a million tonnes of the explosive Trinitrotoluene (TNT). The previous recorded major meteorite impact in 1908 was at Tunguska, which was estimated to cause an explosion of 10 to 20 million tonnes of TNT. These would have been real city killers had they landed in the wrong place. Going further back in time, we had the Meteor Crater impact in Arizona, with an explosion of around 50 million tonnes of TNT, and the Dinosaur-killing Chicxulub extinction event, estimated at 60 billion tonnes of TNT. To be able to deflect any one of this wide range of meteorite sizes, we need a kinetic impactor to deliver the correct kinetic energy, and the deep space probe can provide this variation.

To maximise terminal velocity and thus kinetic energy, we can consider a ground-launched probe with zero payload mass. Figure 12.10 shows the kinetic energy that can be delivered on impact with the asteroid, depending on the propulsion period used. To give a feel for these energy levels, the maximum propulsion period will result in an impact energy of 1.15×10^9 megaJoules, equivalent to 273 thousand tonnes of TNT. For comparison, the Hiroshima nuclear weapon yield was around 16 thousand tonnes of TNT. Clearly, the deep-space probe could provide an effective and rapid response to any incoming asteroid threat, and the required impact can be adjusted by varying the propulsion period. The ability to launch a number of these probes with increasing kinetic energy impacts will give a better chance of obtaining the correct deflection, avoiding the drama sometimes predicted in the media. It is of particular interest that a 2017 TV series called *Salvation* [3] introduced an EM Drive solution to their asteroid deflection problem.

Science missions to orbit each planet in our solar system, or indeed to land on them or their moons, can be carried out by DSPs. An image of a DSP approaching Jupiter is given in Figure 12.11. The maximum and minimum ranges to each planet are given in Table 12.3, together with the flight times for the mean distances.

To travel beyond our solar system to the nearest star requires a different approach, where total loss LH2 cooling is replaced by a closed-loop refrigeration system. This was originally proposed in the International Astronautical Congress (IAC) 2014 paper [4], where an interstellar probe was described. The engine design has now been updated to a fixed L-band engine, still cooled by liquid nitrogen. This gives cavity operation at

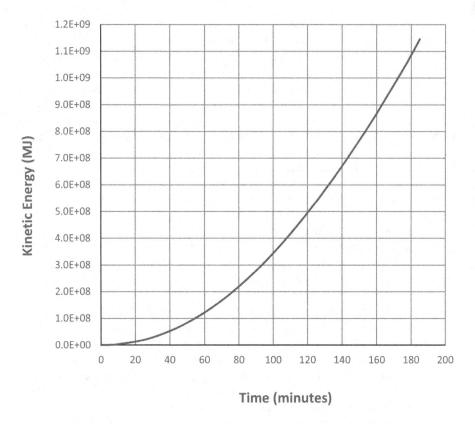

FIGURE 12.10   Kinetic energy for an asteroid deflection mission.

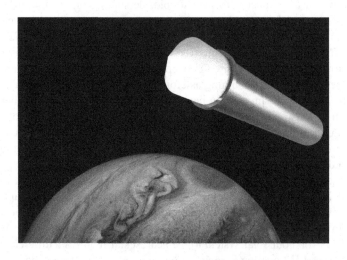

FIGURE 12.11   A deep space probe approaching Jupiter (Jung Dong Whi).

78°K and therefore reduces the refrigeration requirements. A nominal 304 N/kW specific thrust is predicted for a maximum acceleration of 1 m/s/s. A 10-year propulsion period for the Interstellar Probe is defined by the design of the nuclear reactor and electrical generating system. This is based on the work described in [5].

A 1.2 MW thermally rated reactor, using a direct Brayton cycle design operating at 1,300°K, will provide 200 kW of electrical power, with a high-temperature radiator operating between 800°K and 505°K. This is coupled to the propulsion system by a truss structure. The propulsion system comprises the Interstellar Probe (ISP) main engine with its refrigeration module and a low-temperature radiator operating between 280°K and 78°K. The refrigeration module is a reverse Brayton cycle with two-stage turbines. The complete spacecraft is illustrated in Figure 12.12 and has overall dimensions of 28.2 m in length and 12.8 m in width across the high-temperature radiator. The mass budget is given in Table 12.3.

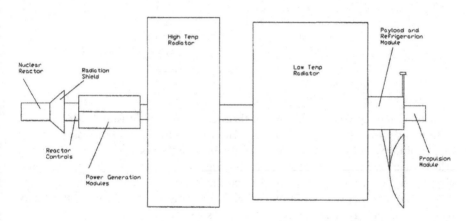

FIGURE 12.12   Interstellar probe.

TABLE 12.3
DSP Missions to Solar System Planets

Planet	Max Range (million km)	Min Range (million km)	Mean Distance (million km)	Flight Time (days)
Mercury	216	83	149	4.8
Venus	260	40	150	4.8
Mars	400	56	228	7.3
Jupiter	966	592	779	24.9
Saturn	1,652	1,204	1,428	45.7
Uranus	3,155	2,587	2,871	91.7
Neptune	4,685	4,311	4,498	143.8
Pluto	7,524	4,293	5,908	188.8

TABLE 12.5

Mass Budget for Interstellar Probe

Item	Mass (kg)
Reactor	528
Shield	600
Reactor controls	31
Heat exchangers	366
Generators	1,612
SSPA	280
High-temperature radiator	885
Main engine	600
Attitude engines	90
Cooling turbine	510
Low-temperature radiator	1,314
Structure	1,400
Payload	720
Total	8,936

With an input power of 29.6 kW, the thruster produces a thrust of 8,998 N, which provides an initial acceleration of 1 m/s/s. However, as the velocity increases towards the terminal velocity, which is close to two-thirds of the speed of light, relativity effects have to be taken into account. To investigate the effect of relativistic velocities on the thrust of an EmDrive engine, first consider a cavity at rest.

Assume a full power wavefront travels a total distance of D inside the cavity before all the energy is dissipated in wall losses, where:

$$D = 2Q_u L \tag{12.1}$$

where Q_u = Unloaded Q value.
    L = Cavity length.

The time to travel the distance D is defined as the time constant of the cavity T_c, where:

$$T_c = \frac{Q_u}{F_0} \tag{12.2}$$

Next, assume that at cavity velocity V_1, the cavity travels a distance of x during the time T_c.

From a Newtonian viewpoint, during this period, the wavefront would have travelled a total distance of $D + x$. However, special relativity theory states that the velocity of propagation of the wavefront is independent of the velocity of the cavity; therefore, the total distance travelled by the wavefront must remain constant.

Thus, as the cavity velocity increases, the distance travelled by the wavefront inside the cavity D must decrease. For D to decrease, Q_u must decrease.

Thus, from equation (12.1), at velocity V_1:

$D = 2Q_1L + x$, where $Q_1 < Q_u$

Then

$$2Q_1L + x = 2Q_uL \qquad (12.3)$$

Now $x = V_1T_c$

Therefore, from equations (12.2) and (12.3):

$$2Q_1L + V_1\frac{Q_u}{F_0} = 2Q_uL$$

Then solving for Q_1:

$$Q_1 = Q_u\left\{1 - \frac{V_1}{2LF_0}\right\} \qquad (12.4)$$

Applying the modified value of Q from equation (12.4), the interstellar probe mission profile is given in Figure 12.13.

Thus, after 9.86 years of propulsion, the velocity has reached 204,429 km/s at a distance of 3.96 light years. It is anticipated that following the propulsion period, all electrical power would be utilised by the communications system during the subsequent target fly-by. The probe would be launched into Earth's orbit from the heavy launch vehicle, to be described in the next chapter.

The deep space and interstellar probes, which have been described, should raise interest among space scientists and astronomers, but they do not guarantee that the technology will flourish. Indeed, the academic world in particular seems very slow to understand EmDrive, even though the physics is pretty elementary. Perhaps the engineering challenges are just too difficult, or perhaps the peer pressure to conform to the groupthink that "EmDrive is impossible", still prevails. The controversy that EmDrive caused is illustrated by the large number of posts that appeared on a NASA spaceflight forum over a period of 7 years [6]. Apparently, the forum received over 6 million reads, covering 12 separate threads. Thus, the narrow section of the population comprising academics and their enthusiastic online followers is very slow to accept EmDrive. The only people who have apparently shown progress are those in the intelligence communities, as illustrated by the early interest by the National Security Agency (NSA) and other similar agencies, together with the extraordinary leaks of UAP sightings and their follow-up investigations. However, the next two vehicles should excite a wider public, as they offer solutions to our upcoming and seemingly inevitable global warming crisis. Also, a revolution in personal and commercial transportation, both on Earth and in space, can be confidently predicted using these vehicles.

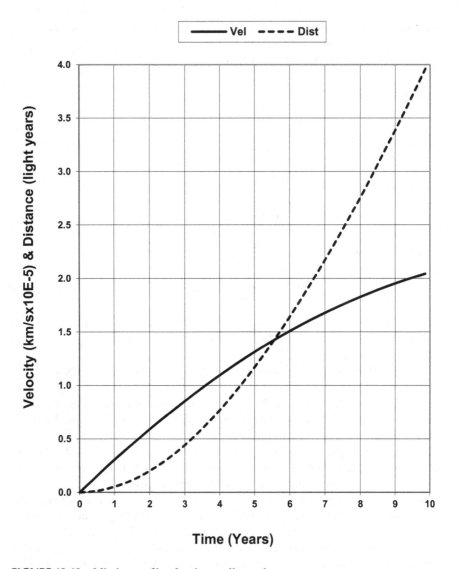

FIGURE 12.13 Mission profile of an interstellar probe.

REFERENCES

1. Ohndorf A et al. IEPC-2011-051. IEPC conference Wiesbaden September 2011.
2. Elvis M. *Asteroids*. Yayle University Press 2021. ISBN 978-0-300-23192-2.
3. Salvation. American suspense drama TV series. 26 episodes over 2 series. July 2017 to September 2018.
4. Shawyer R. Second Generation EmDrive Propulsion Applied to SSTO Launcher and Interstellar Probe. *Acta Astronautica*.116 (2015) 166–174.
5. Blott, Donaldson and Mazzini. High power nuclear electric propulsion. IEPC-2011-061 Wiesbaden, 2011.
6. NASA spaceflight forum. EM Drive Developments Thread 1 (nasaspaceflight.com)

13 A Heavy Launch Vehicle

For a radically new technology to be adopted, there is inevitably an urgent, important requirement to be met. History tells us that this very often occurs in the lead-up to, or indeed during major conflicts. Of the three main driving forces for innovation, namely fear, greed or curiosity, fear is dominant, as the defence budgets of most countries will illustrate. However, conflict between states is negligible compared to the future conflict between humanity and the planet. A burgeoning population, with a rapidly increasing energy demand, continues to use fossil fuels to generate that energy. Scientists are now agreeing that the burning of fossil fuels has increased the concentration of carbon dioxide in the atmosphere, which is leading to a runaway increase in global temperature and disaster for the human race. There is clearly a need to develop alternative energy sources. The easiest approach is to harness wind and ground-based solar technology, although these have obvious limitations. They do not work when the wind doesn't blow and the sun doesn't shine. Energy storage technologies using batteries, water storage or molten salt offer limited solutions, although the installations required for a few days' energy provision become enormous. Political opposition to nuclear fission and the problems of nuclear waste continue to plague that route, and the hopes for nuclear fusion always seem thirty years in the future.

Although the use of space-based solar energy generation using solar power satellites (SPS) may seem like science fiction, the basic technology is well understood, and it is simply a matter of implementation. Transmission of the electrical energy via microwave beam does not present any great technical or safety challenges. The receiving stations will certainly cover a significant area, with elliptical diameters of several kilometres, but for a given power output, they are much smaller than ground-based solar arrays. The seminal study, in the IAA report of November 2011 [1], concluded that there are no fundamental technical barriers that would prevent the realisation of large-scale SPS platforms during future decades; however, questions remain as to their economic viability. A major conclusion of the report was that the most critical challenge is the essential requirement for extremely low-cost Earth-to-Orbit transport (EOT). Also, acceptable EOT systems for future SPS must be environmentally benign, i.e., space transportation infrastructures to launch the satellites cannot result in harmful pollution of the atmosphere. The current approach to this challenge is to develop large reusable conventional rockets, which will only enable a partial solution to the requirement. Thus, since the SPS solution was identified in 2011, the global warming problem has become worse, and the main challenge of extremely low-cost EOT has not been adequately addressed. We are running out of time.

A useful way of illustrating the challenge is to estimate the specific cost of launch in dollars per kilogram of payload to reach geostationary orbit (GEO). The lowest mass estimate for a 2 GW SPS in 2011 was for the Type 111 highly modular satellite,

DOI: 10.1201/9781003456759-13

with a total in-orbit mass of 6,700 tonnes. Although a number of further studies have been carried out with different satellite architectures, significant mass reductions have not been achieved. For conventional launch vehicles available at the time, a total of 1,700 Atlas V launches would have been required at a total cost of $194 billion. This equates to a specific launch cost of $29,000 per kg. The study suggested an order of magnitude reduction was required down to a total cost of $20 billion, or $3,000 per kg, before space-based solar power could even begin to be viable. From the latest estimates available for a proposed large reusable conventional rocket [2], a payload mass to geotransfer orbit of 21 tonnes is offered. To circularise the orbit, a further 1.8 km/s increase in velocity is required, which, from typical GEO communication satellite design, requires at least half the mass to be an apogee propulsion system. Thus, for a single launch of say 10 tonnes to GEO and a projected cost of $10 million per launch [3], a specific cost of $1,000 per kg is the best estimate currently available for a reusable conventional rocket launch to GEO. We therefore have a potentially viable, SPS-based solution to the global energy requirement. Accordingly, a number of further international studies into space-based solar energy technologies were started, and at the 2021 IAC conference in Dubai, we presented a paper outlining the potential cost savings of using an EmDrive-based heavy launch vehicle [4]. The resulting specific cost of $10.9 per kg, almost two orders of magnitude lower than the best conventional propulsion cost estimate, makes the economic case for SPS launched by EmDrive unassailable in the future.

So what would a heavy launch vehicle, designed for the SPS application look like? From our early work on the EmDrive spaceplane design, illustrated in Figure 7.8, we knew that an airframe design similar to the aerodynamic test model would fly. Also, later discussions, particularly with the USAF, after our Pentagon meeting in 2008, as described in Chapter 7, resulted in an outline requirement specification. Basically, the uncrewed vehicle should be capable of delivering a minimum of 50 tonnes of payload to GEO. It would need to be reusable, with a 500-mission lifetime as a minimum. Vertical take-off and landing with emergency glide and horizontal landing would be required. The payload should be capable of being carried un-faired, thus limiting the in-atmosphere velocity. In orbit, manoeuvrability should be good to assist in the assembly of the SPS components once the designated orbital slot has been reached. All this needed to be achieved at a low vehicle cost, implying a manufacturing concept similar to a typical jet airliner. The low-cost approach also needed to cover the vehicle operations concept, with a fast turn-around time on the ground and minimum launch site facilities. The latter requirement, together with vertical take-off and landing operation, would enable vehicle operations, i.e., loading, fueling, launching and landing, to take place at the SPS component manufacturing site. This would minimise the ground transport requirements for the SPS components.

The resulting design is illustrated in Figure 13.1, where a payload comprising four SPS roll-up arrays is illustrated. The payload is hung beneath the vehicle from a sub-frame with attachment rails and an undercarriage tailored to the specific payload. This would enable large, complex SPS components, including solar arrays and truss assemblies, to be carried. With the unfaired payload components carried in this way, they would be easily loaded up on the ground and directly attached to the SPS

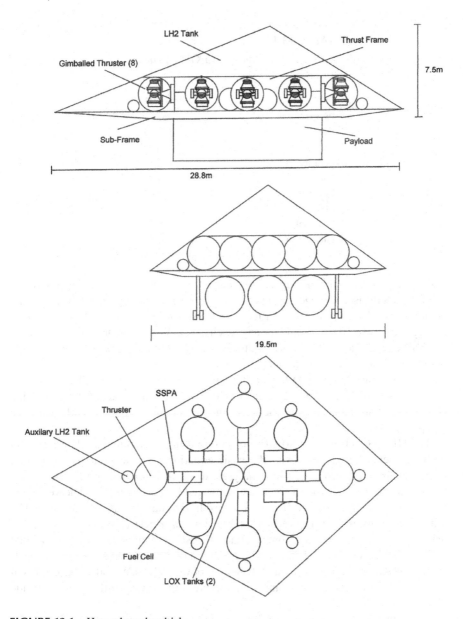

FIGURE 13.1 Heavy launch vehicle.

in orbit. The HLV thrusters could even be used to roll out a solar array as part of the robotic assembly of the SPS.

The HLV comprises a large LH2 tank mounted on the top of a thrust frame containing LOX tanks and eight fully gimballed thrusters. The thrusters are similar to those used in the Deep Space Probe described in Chapter 12. The LH2 tank comprises two separate tanks for redundancy and 30 cm thick insulation to minimise

TABLE 13.1
HLV Mass Budget for a GEO Launch
Mission

Item	Mass (tonnes)
Payload	65
Fuel	7.59
Airframe	14.33
Thrusters	5.12
SSPAs	11.97
Fuel cells	8.27
LH2 tank	3.53
LOX tank	0.99
Launch mass	116.8

boil-off during ground handling and extended duration in space. The SSPA and fuel cell for each thruster are packaged and arranged to fit within the thrust frame, while an auxiliary LH2 tank is attached to each thruster to maintain cavity temperature during maintenance periods and for engine changes. As the vehicle is not crew-rated, the SSPAs and fuel cells are not redundant, although full vehicle control can be maintained with four of the eight engines out.

A mass budget for a GEO launch mission is given in Table 13.1. The mass and dimensions of the SSPAs and fuel cells, which form a major part of the propulsion system mass, are based on specific mass and volume data from currently available equipment.

The HLV launch mission to GEO is analysed in a different way than a typical ballistic rocket launch; however, the basic equations are the same as those used in the Deep Space Probe mission analyses, described in Chapter 12. The launch mass is overcome by a direct lift force generated by four thrusters. The vehicle is therefore vertically launched from a standard horizontal aircraft-like attitude. The other four thrusters provide vertical and horizontal acceleration forces as well as primary pitch, roll and yaw control. Each thruster, complete with its associated SSPA and fuel cell, is rated at a maximum thrust of 320 kN and a maximum acceleration of 9.81 m/s/s for an input microwave power of 80 kW and a DC electrical power of 178 kW. This allows any four thrusters to provide full lift and attitude control for the vehicle.

The numerical analysis is therefore carried out by calculating velocity on the horizontal axis and rate of climb and altitude on the vertical axis for increments in mission time. The launch objectives are to achieve an orbital velocity (horizontal) of 3,075 m/s at an altitude of 36,000 km with a final rate of climb of zero. The launch mass of 116.8 tonnes is reduced during the flight as the fuel mass decreases. LH2 is initially used for thruster cooling, and the rate of LH2 use is dependent on the total microwave power input to the thrusters at any given point in the flight and the latent heat of LH2. The flow rate of LOX is dependent on the DC electrical output of the fuel cells, while the H2 input is some of the boiled-off gas from the thruster cooling

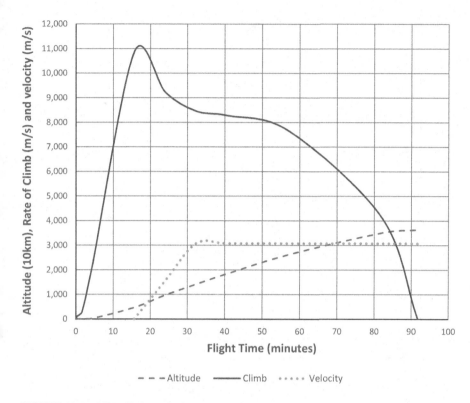

FIGURE 13.2   HLV flight to GEO.

process. The remainder of the cold H2 gas is used for SSPA and fuel cell cooling before being vented from the top of the vehicle. Clearly, the analysis needs to be run a number of times in an iterative process to achieve the launch objectives while optimising total fuel use. Figure 13.2 shows the mission envelope for a 92-minute flight to deliver a payload of 65 tonnes to GEO.

The initial rate of climb, taking no more than 3 minutes and reaching 100 km altitude, is restricted by throttling the thrusters down to a minimum of 50% thrust. This limits the atmospheric drag caused by the pyramidal shape of the LH2 tank. The tank provides a shield to protect the payload from atmospheric forces and thus removes the need for a payload fairing. Once clear of the atmosphere, the vertical climb then rises to a maximum of 11 km/s until, at a flight time of 16 minutes, the thrusters are rotated to a horizontal attitude to commence acceleration to orbital velocity. The 3 km/s orbital velocity is achieved by a flight time of 32 minutes, allowing the vehicle's climb to continue under the deceleration of the reducing gravitational force. At 82 minutes of flight time, the thrusters are rotated to a vertical upwards attitude to provide a 10 minute period of deceleration, such that the climb rate reaches zero at 36,000 km altitude.

The complex thrust regime produces the vehicle acceleration profile illustrated in Figure 13.3.

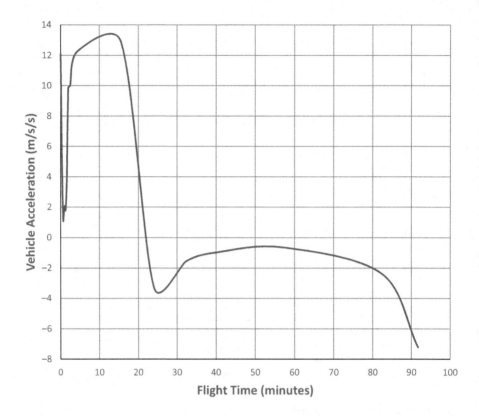

FIGURE 13.3   HLV vertical acceleration profile to GEO.

Figure 13.3 shows the initial vehicle acceleration dropping to just over 1 m/s/s as the thrusters operate to give low acceleration thrust and while drag forces are at maximum. Vertical acceleration then increases to a maximum of 13 m/s/s, until at 16 minutes, it drops to below zero as the thrusters are turned to a horizontal attitude, and vertical acceleration becomes vertical deceleration due to gravity. The final 10 minutes of deceleration due to the downward thrust reach a maximum of 7 m/s/s. The total fuel mass used to reach GEO is 3.3 tonnes, leaving ample fuel margin for orbital manoeuvres, LH2 boil-off, and the return flight. Fuel is also required to move the vehicle at low altitude from its initial launch location to the designated area for the vertical ascent. This allows flexibility over the loading and initial launch locations while meeting any safety requirements for the area under the main ascent.

The dominant component of the overall operational cost of the SPS launch is the initial development cost of the HLV. It has always been the objective to operate the vehicle for a minimum of 500 missions before major maintenance or replacement. For an uncrewed vehicle with a fully solid-state propulsion system and low mechanical and thermal stress during flight, this is considered a conservative approach. The wide range of space and terrestrial transport applications will ensure that large numbers of vehicles will be built. The low acceleration and low atmospheric velocities result in low mechanical and thermal stress on the airframe. This means that conservative airframe design and conventional materials will lead to mass production techniques more familiar to the truck industry rather than the aerospace industry. This approach will ensure that costs are reduced significantly compared to traditional aerospace programmes.

It is therefore estimated that the unit cost will be approximately $250 million. This may be compared with the current quoted price of $185 million for a 767–300F freighter aircraft, which has a similar payload capacity but a higher take-off mass and fuel load. For 500 missions, the capital cost is therefore $500k per mission. This assumes no write-off costs per vehicle. This is conservative, as following a refurbishment, a second ownership sale will bring in additional revenue. The main operating systems on the vehicle (control and propulsion) are all solid-state, and with the exception of cryogenic valves and thruster gimbals, they are unlikely to require significant maintenance over 500 missions. It is assumed that with a total flight time of less than 3 hours per mission, efficient on-orbit operations, and rapid ground refuelling and payload attachment, one mission per day would become routine. Assuming LH2 costs $0.7 per kg for manufacture at the launch site and LOX costs $0.1 per kg, the total fuel cost per mission is $12.9k. Assuming a shift team of 20, with each mission requiring three shifts, labour and overhead costs are estimated at $31.2k. Note that the launch site consists of an area of tarmac smaller than a supermarket car park, with an adjacent hanger for payload attachment and refuelling. Pre-flight checkout will be largely automated, as are most flight operations, with manual remote control only required for SPS docking and payload deployment. All ground operations would be modelled on routine airline practice. The total cost per mission is therefore estimated at $544.1k. For a 65-tonne payload, the launch cost to GEO is therefore $8.4 per kg. This dramatic reduction in launch costs would make SPS the prime choice for renewable energy supply and thus the primary solution to the ever-closer global warming crisis.

Once certification for flights between ground locations has been obtained, the use of long-haul, point-to-point cargo transport could evolve from the GEO launch application. Figure 13.4 shows the HLV carrying four standard-sized shipping containers. In this application, the vehicle would climb to above 100 km altitude and be able to transport and land a cargo of 67 tonnes anywhere in the world, i.e., a maximum range of 20,000 km. The mission analysis gives a flight time of 48 minutes, with a maximum velocity of 16 km/s and acceleration levels limited to below 1.3 g. A fuel margin of 7.5% would be left, and the ability to optimise between range, payload mass, and fuel mass would enable a wide variety of one-way or return missions to be planned.

To further illustrate the versatility of the HLV design, an example of an exciting deep space science mission can be envisaged if ten of the EmDrive-propelled Deep Space Probes, described in Chapter 12, are launched into space. The payload carrier, mounted under the vehicle sub-frame, is shown in Figure 13.5.

By launching from a position closer to the sun, the ten probes can fan out to give two orthogonal patterns of probes at 60 degrees of angular separation. This would enable simultaneous field and particle measurements of the whole volume of the solar system, from close to the centre to the furthest edge of the heliopause. An additional LH2 tank would be carried in the nose of the payload carrier to top up the probe tanks during the flight to the launch point. The payload carrier would also provide

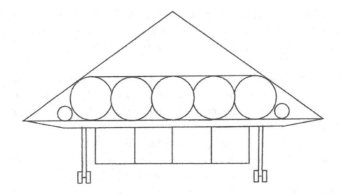

FIGURE 13.4 HLV with a payload of four shipping containers.

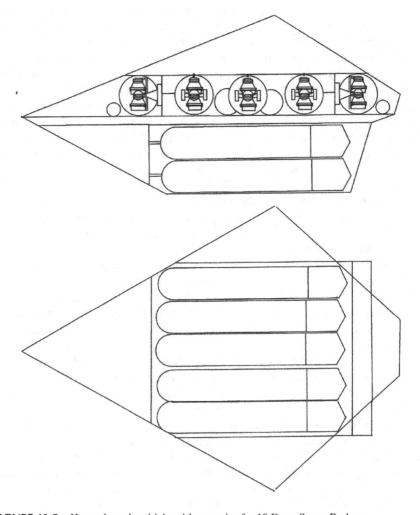

FIGURE 13.5 Heavy launch vehicle with a carrier for 10 Deep Space Probes.

the undercarriage for the vehicle and would increase the total payload mass from 50 tonnes for the 10 fully fuelled probes to an estimated 60 tonnes.

A further application would be to use the HLV in support of crewed missions to the Moon or Mars. A Mars cargo mission analysis concludes that, by utilising the full fuel capacity of the HLV (22.6 tonnes), 50 tonnes of cargo can be landed on Mars in a minimum of 9.5 days. A typical cargo would be a habitation module, shown in Figure 13.6, which would be landed in advance of a crewed mission. This analysis assumes a minimum range to Mars of 56 million km, which occurs every 2 years. At the maximum range of 401 million km, the flight time would be 68 days, with the additional time taken in the cruise phase when the thrusters are switched off. Launch would follow the 10-probe HLV launch profile, with acceleration and deceleration levels up to a maximum of 2.4 g, but with a longer cruise period. A maximum cruise velocity of 68 km/s would be achieved.

The precision manoeuvrability of the vehicle would be utilised in the landing phase of the HLV, when the habitation module would be positioned and docked with any existing module while still hung from the HLV subframe. The retractable landing skids mounted on the module would then be lowered to support and level the module on the Martian surface. This use of the HLV as a skycrane, with the necessary six degrees of freedom, would enable the assembly of a multi-module Martian base with the minimum need to prepare the area.

A commercially attractive application of the HLV would be the delivery of mining and precious metal extraction equipment to any asteroid that was shown

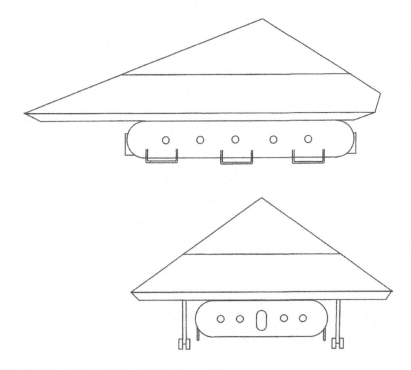

FIGURE 13.6 HLV with Mars habitation module.

to be suitable by the DSP prospecting missions, described in Chapter 12. The HLV would then be used to transport the cargo, worth a minimum of $1 billion, back to Earth.

REFERENCES

1. Mankins J. *Space Solar Power.* International Academy of Astronautics. 2011.
2. Starship Users Guide. Space X. Revision 1. March 2020.
3. Elon Musk says he's 'highly confident' that SpaceX's Starship rocket launches will cost less than $10 million within 2-3 years. www.business insider.com 11 Feb 2022.
4. Shawyer R. *The impact of EmDrive Propulsion on the launch costs for Solar Power Satellites.* IAC-21,C3,1,5,x62540.

14 A Personal Spaceplane

In the 2019 IAC paper [1], a moon landing and return mission were described using the 3G technology described in Chapter 10. The mission was the result of a study carried out for Pascal, a French colleague who established Soft Industry Development, a company based in Seoul, Korea, developing animation and theme park concepts. Pascal's interest in blending arts and sciences led to the 2014 launch of the Luna project, consisting initially of producing promotional material to celebrate the up-coming 50th anniversary of Apollo 11. The subsequent concept of establishing a space tourism industry to explore the moon meant that the cost and danger of Apollo-type missions had to be overcome. EmDrive propulsion offered a solution. With the evolution of 4G technology, the acceleration levels for a moon mission could be increased, although for tourist missions these would be kept to a comfortable 1 g. This enabled the flight time from Earth to the Moon to be reduced to less than 5 hours, compared to the 37 hours of the original mission plan. The vehicle crew capacity was also increased from 3 to 6, and the resulting vehicle design became the personal spaceplane (PSP), illustrated in Figure 14.1.

The PSP is one of a family of vehicles based on a four-engine, thrust frame and weighing up to 22 tonnes. The orbital version, which could also be used for the moon landing mission, is detailed in Figure 14.2. This version would include a 6-person cabin with full life support equipment and a passive docking hatch to the rear, which is compatible with the NASA International Docking Adapter-2. It is envisaged that the six seats can be folded down to give a volume large enough for passengers to enjoy a zero-gravity experience during the cruise part of their flight.

The airframe has upper and lower elevons on the rear outer edges of the wings. These four control surfaces can achieve pitch, roll and yaw control, as well as being

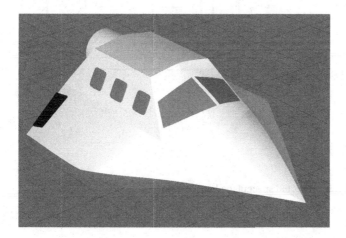

FIGURE 14.1 External view of the personal spaceplane design (Jung Dong Whi).

DOI: 10.1201/9781003456759-14

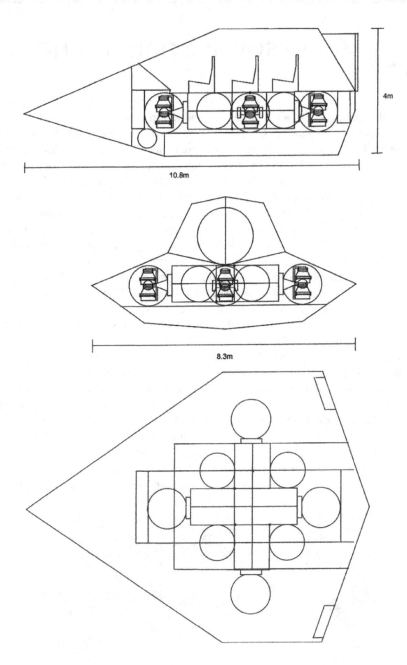

FIGURE 14.2 Personal spaceplane layout.

used as air brakes. The spaceplane is designed to glide in an all-engine-out emergency. In this event, a fast helium purge system would empty the front and main liquid hydrogen tanks and give the vehicle high crash worthiness. The empty tanks, with internal bladders, would also provide good buoyancy for a crash landing on water.

The EmDrive engine used in the PSP and shown in Figure 14.3 is the same as the one used in the Heavy Launch Vehicle, but with a lower nominal power rating of 48 kW and certified for commercial crewed space vehicles. To obtain full space qualification for a crewed spacecraft, the spacecraft needs multiple operating modes to cover mission emergencies. The four-engine configuration, with the interconnection of power amplifiers and fuel cells, gives full redundancy. The maximum thruster power rating of 80 kW enables an emergency vertical landing to be undertaken with only one engine functional, albeit in a non-normal vehicle attitude.

Thus, as well as having dual cavities, primarily for Doppler correction, all major components have dual redundancy. A larger auxiliary LH2 tank is attached to the side of one of the fuel cells via a thermal interface. This provides emergency LH2 supply to the engines and enables the cavities to be maintained at LH2 temperature during routine maintenance periods. The engine is mounted on an L-shaped plate capable of transferring forces to the space-plane thrust frame in all directions. The LH2 input and the H2 output for the cavity coolers are via rotatable joints in both sets of gimbals. Because of the inevitable leakage at these joints, the whole thruster assembly is contained in a stainless steel spherical casing. A small vacuum pump is used to scavenge the H2 gas from the casing. The microwave input to the isolators at the cavity inputs is via the second gimbal joint and sections of waveguide. The thruster casing provides isolation from radiated microwave leakage from the joints. The dual fuel cells and solid-state power amplifiers (SSPAs) are enclosed in standard space-construction aluminium boxes, machined for minimum mass.

Any mission emergency in space that required crew and passenger escape would use a high-performance version of the spaceplane with two additional seats and higher acceleration levels to enable rapid rendezvous and rescue via the docking adaptor. The HLV engine power level of 80 kW would be adopted, together with an extended LH2 tank in the nose. The high levels of manoeuvrability given by the four

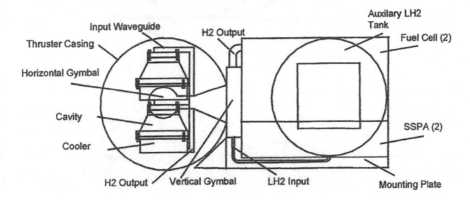

FIGURE 14.3   Personal spaceplane engine.

2-axis gimballed engines would enable fast approach and docking procedures. The airframe of the rescue spaceplane would be stiffened and strengthened to withstand up to 9 g acceleration forces. All equipment would also be qualified for this acceleration level. The rescue crew would wear military-type anti-g suits and would be fully trained in high-acceleration flight and docking manoeuvres. A drawing of the rescue spaceplane is given in Figure 14.4, showing the outline of the modified airframe.

It is worth noting that this upgraded rescue spaceplane would make a formidable aerobatic aircraft. The four fully gimballed thrusters provide six degrees of freedom, i.e., pitch, roll and yaw, together with translation in the X, Y and Z planes. This gives the spaceplane vectoring in flight capability. High-speed flight within the atmosphere would be enabled by the more aerodynamic profile while retaining the generic internal layout of the standard PSP.

A mission analysis has been carried out for a manned moon landing and return. A payload mass of 5,000 kg has been assumed to cover the cabin, passengers and life support equipment. The total vehicle dry mass plus payload is 18,148 kg, and the maximum fuel mass is 2,818 kg leading to a launch mass of 20,966 kg. The vehicle attitude at launch is horizontal, with acceleration in the vertical direction. A 90-degree rotation is then made to give a vertical acceleration with minimum air drag. The acceleration levels referenced to the local vertical are shown for Earth, Moon and vehicle in Figure 14.5.

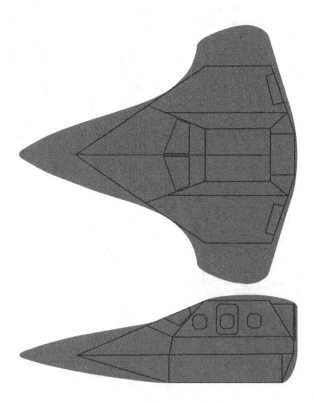

FIGURE 14.4   Outline of a high-performance personal spaceplane (Jung Dong Whi).

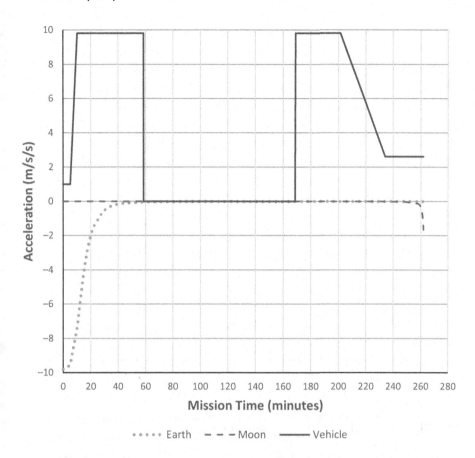

FIGURE 14.5 Acceleration levels for the local vertical axis.

An acceleration of 1 m/s/s is maintained until an altitude of 100 km is reached. This minimises drag, which is a maximum of 2,161 kg at an altitude of 10 km and a velocity of 100 m/s. Once clear of the atmosphere, acceleration increases to 9.81 m/s/s (1 g) until a distance of 150,000 km from Earth is reached. At this point, the acceleration drops to zero, and the vehicle is in cruise mode. This flight path optimises fuel usage. A 180-degree rotation is made to align the vehicle for the deceleration phase. The cruise lasts for 96 minutes, and during this period, the passengers will experience zero gravity. At 310,000 km from Earth, deceleration at 9.81 m/s/s will start, reducing to 2.61 m/s/s at 360,000 km from Earth. A vertical landing is carried out in the same horizontal vehicle attitude as at launch, at a distance from Earth of 384,415 km, after a total flight time of 262 minutes (4 hours, 22 minutes).

Vehicle velocity and fuel usage are illustrated in Figures 14.6 and 14.7. Total liquid hydrogen used for engine cooling is 1,022 kg, while liquid oxygen, used for power generation together with boiled-off hydrogen gas, is 287 kg. For simplicity, the total mission fuel used is assumed to be twice the Earth-to-Moon flight. Thus, a conservative fuel margin for the full mission is 8.5%.

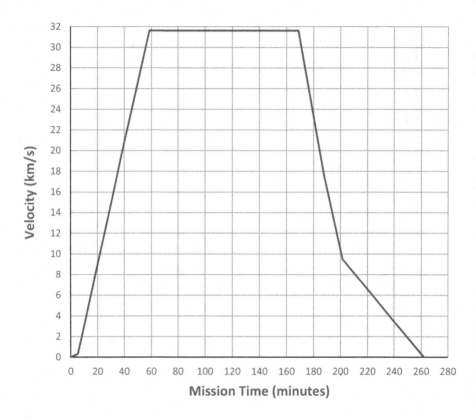

FIGURE 14.6   Vehicle velocity for Earth-to-Moon flight.

The overall dimensions of the PSP cabin shown in Figure 14.2 are 5.5 × 3.9 m and a height of 1.9 m. This gives a relatively spacious internal cabin, allowing for six reclining seats, a hygiene compartment, and spacesuit stowage, together with controls and life support equipment. A typical crew for a tourist flight to the moon would consist of one company pilot, one flight attendant, and four paying passengers. They would be transported in reasonable comfort to and from habitation modules on the moon's surface, which would form a moon base for scientific and tourism purposes. It is envisaged that for paying passengers, training would consist of a medical, followed by a few days of PSP familiarisation and safety procedures. Spacesuit fitting and training would be necessary, as a moonwalk would become the "must do" part of a tourist trip as well as a useful safety precaution. Finally, rescue training from the spaceplane would be mandatory, which would include recovery from a simulated emergency landing on water.

The ground facilities for a fleet of five PSPs would comprise two 11,000-square-foot buildings. The buildings would be one flight hanger and one service hanger. The flight hanger would accommodate up to three PSPs and be used for fueling, loading and first-line maintenance. Due to the presence of liquid hydrogen and liquid oxygen, this building would be classified as a hazardous area and be subject to special safety conditions. The service hanger would enable routine servicing to be carried out on

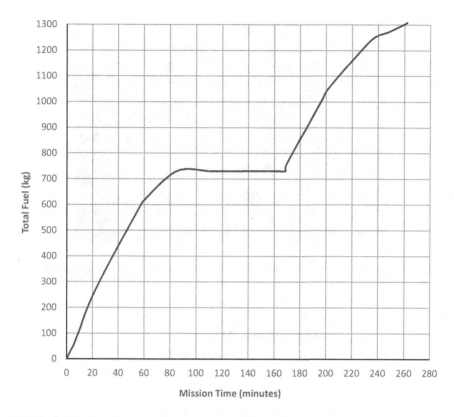

FIGURE 14.7   Total fuel used for the Earth-to-Moon flight

up to three PSPs and accommodate mission control and staff facilities. As launches and landings are purely vertical, the actual launch and landing site would be no more than a small area of tarmac close to the flight hanger. The only reason to site the facility at an existing airport or spaceport would be the need to use controlled airspace. Figure 14.8 shows the PSP coming in for a vertical landing at the Cornwall Spaceport, UK.

With modifications to the cabin for a three-person crew, and provisions for longer flight times, the PSP can be used for crewed missions to Mars. In Chapter 13, we saw that the flight time of an uncrewed Heavy Launch Vehicle (HLV) cargo mission to Mars could be as low as 9.5 days. However, this would increase to 68 days when the Mars-to-Earth distance is at its maximum. This needs to be reduced for a crewed mission. The HLV mission also assumed a maximum acceleration of 2.4 g. The effect on the human body of sustained periods at these acceleration levels is not yet fully understood, and therefore the crewed Mars mission analysis limited acceleration to 1.2 g. The first mission would be carried out at minimum range and have a flight time of 11 days. Importantly, the analysis gives sufficient fuel margin for an immediate return flight if there are problems docking with the habitation module, which would have been prepositioned on the Mars surface by the first HLV Mars mission. It is envisioned that a second HLV mission would carry a fuel store of 50 tonnes of LH2

FIGURE 14.8   A personal spaceplane approaches Cornwall Spaceport (Jung Dong Whi).

TABLE 14.1
Sequence of Mars Missions

Mission	Payload
HLV 1	Habitation module 1
HLV 2	Fuel store
PSP 1	3 crew
HLV 3	Logistics module and Node 1
HLV 4	Habitation module 2
PSP 2	3 crew
HLV 5	Solar power satellite
HLV 6	Rectenna components
HLV 7	Aeroponic module and Node 2
HLV 8	Electrolyser and ice mining equipment
HLV 9	Science module

and LOX. This would enable the crew to refuel the PSP and return to Earth at any point in the 2-year cycle to the next minimum range. With full fuel tanks, the PSP can cover the maximum range in 43 days while still keeping acceleration below 1.2 g and providing an acceptable flight time for any emergency return. It is envisaged that normal Mars base operations would involve long periods for the crew of up to 2 years. The objective would be that during the first period, the base would build up to become self-sufficient for a crew of 6. This would involve a sequence of HLV and PSP missions as given in Table 14.1, and an outline diagram of the resulting Mars base is given in Figure 14.9.

HLV 1 and HLV 2 will first land close to each other under the autonomous control system of the HLV. Note that the communications time delay between Mars and Earth of between 5 and 20 minutes one-way means that all vehicle piloting would

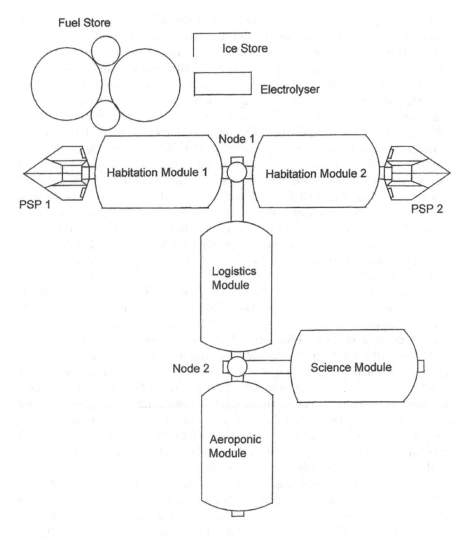

FIGURE 14.9   Mars base.

either be autonomous or under local crew control. PSP 1 will then land and dock with Habitation Module 1 under the control of the PSP pilot. With the PSP 1 crew safely aboard Habitation Module 1, HLV 3 can land. Node 1, with the logistics module attached, can then be docked to the other port of Habitation Module 1, with the HLV in sky crane mode being controlled locally by a PSP 1 crew member. HLV 4 would land next and dock Habitation Module 2 at the third port of Node 1. This initial base would provide minimum habitation for a crew of six and therefore enable PSP 2 to land with a further three crew members. The Mars base crew would be made up of professional astronauts, with differing skill sets and could well comprise two pilots, an engineer and a medic, and two geologists. The initial six missions would be planned to be completed within 2 months around the period of minimum range, thus minimising flight times and allowing two HLVs to complete two missions each. The remaining five HLV missions would enable the base to be completed.

The HLV 5 to HLV 9 missions would be spaced out during the 2-year cycle according to the work rate of the crew. HLV 5 will insert a 50-tonne solar power satellite (SPS) into an aerostationary orbit around Mars. This orbit, 17,032 km above the surface of Mars, will enable solar energy to be beamed to a rectenna array via a microwave beam controlled by a pilot signal from the Mars base. Scaling from the SPS design, to be described in Chapter 15, and allowing for the lower insolation level at Mars, the Mars base could receive up to 15 MW of continuous electrical power. The SPS would also carry a payload to provide a full-time communications link between the Mars base and Earth. Two separate communications satellites would also be carried on board HLV 5 and deployed at 120-degree orbital positions to provide links across almost all of the surface of Mars. This communications system would allow the PSPs to remain in contact with both the base and Earth, during subsequent exploration missions. The rectenna array would be carried on HLV6, and the erection of the large array of posts and wires would comprise a large proportion of the crew's workload. Undoubtedly, robotic vehicles will simplify construction. With such significant levels of electrical power available, high-energy plant growth techniques would be used to provide food for the crew. This would take place in the Aeroponic module, delivered along with Node 2 on HLV 7. Aeroponic plant growth involves growing plants in a fine mist of tiny water droplets using high-pressure equipment. Although high growth rates with small space requirements have been reported for various plants and vegetables, the high energy requirements have tended to hold back the development of such technology. At the Mars base, a sustainable food source is essential, but energy is plentiful. As with EmDrive, aeroponic research dates back to the 1970s. In the summer of 1976, John Prewer carried out a series of aeroponic experiments near Newport, Isle of Wight, U.K., in which lettuces were grown from seed to maturity in 22 days. Work at NASA has continued up to the present day, including in-orbit plant growth experiments.

After energy supply, the second major requirement for long-term base habitation is a water supply. It is suggested that the base would be positioned close to a source of water ice on the surface of Mars. An example would be at the edge of the Korolov crater at 73° North. This 82-km diameter crater is estimated to contain 2,200 km^3 of water ice.

A robotic ground vehicle carried on the HLV 8 mission would be used to excavate the water ice and return it to the base. An electrolyser, also carried on HLV 8, would then melt the ice, remove the dust, and produce LH2 and LOX to maintain the fuel store. Again, SPS electrical power would be plentiful for this process, and the ability to keep the fuel store topped up would ensure short return flights to Earth. Access to fuel for the PSPs would also enable them to be used for extensive exploration of the Martian surface. The temptation to carry out exploration flights out over the surface for increasingly extended periods would give the crew a real sense of purpose. The SPS is an ideal vehicle for exploration, as it provides habitation, access to the surface, and VTOL with no exhaust plume to disturb the surface to be examined. A final HLV mission would provide a fully fitted-out science laboratory module to enable samples from all over Mars to be analysed on site. I suspect that the geologists on the crew will want to sign on for a second 2-year cycle. However, contingency rock samples would inevitably be containerized, for delivery back to Earth by each returning HLV.

The fuel store would also provide a backup electrical supply for the habitation modules using fuel cells. The electrolyser and fuel cell processes would also ensure a biological barrier for producing drinking water and providing oxygen to the life support systems at the base. The process involves reducing the ice to hydrogen and oxygen for storage as liquefied gas before combining them back into water as a by-product of producing electricity. The overall objective would be to establish a fully self-sustaining base using only sunlight and Martian ice within the first 2-year cycle. A block diagram of the overall system is given in Figure 14.10.

Returning to terrestrial applications, lower-cost versions of the PSP, would provide global passenger and freight transport using sub-orbital flight profiles. Mission analyses have shown that a sub-orbital version of the PSP would attain a useful velocity of approximately 5 km/s within 15 minutes. This would put flight times for global distances at around an hour. These applications would only require a pressurised cabin and a simple aircraft-style door, rather than the docking hatch envisaged for the orbital version. A short-route, air taxi version would be even simpler, require no pressurisation, and only be required to meet general aviation regulations. The velocity of an air taxi version, limited to say 5,000 feet of altitude, would be limited by air drag to a few hundred mph. This would nevertheless serve the regional air transport market without the need for regional airports. The VTOL capability, with no exhaust or downdraft, together with silent operation, would give a genuine door-to-door service.

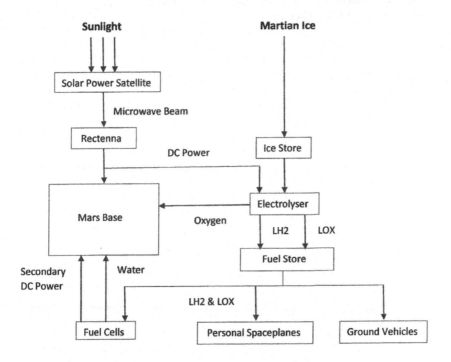

FIGURE 14.10 Block diagram of the Mars base system.

The air taxi version would be particularly useful in an air ambulance or rescue role. The lack of a downdraft or exhaust plume would enable operation in an urban canyon situation, which would otherwise be dangerous for a conventional rotor-driven craft.

The air taxi could also become the prime component of a fully autonomous personal transportation system. The major problem for fully autonomous ground vehicles is the enormously complicated operating environment in which they will need to operate. This environment is also becoming overcrowded, particularly in city areas, where average journey times are getting longer each year. This is fundamentally due to the limitations of two-dimensional travel. Once the third dimension of air is introduced, the overcrowding problem is reduced, and the operating environment is greatly simplified. There are no little old ladies stepping off the pavement into the road at an altitude of a few hundred feet. Clearly, the safety and reliability of these vehicles will need to be demonstrated over a significant period of time before they are generally accepted by the public. However, conventional air travel is now fully accepted, in spite of the inherent risks of high-speed horizontal take-offs and landings on runways. The door-to-door convenience and quiet vertical take-offs and landings will swiftly overcome remaining public concerns for this new form of air travel. A major effect of this potentially huge market would be the reduction of unit costs for the full family of EmDrive vehicles.

REFERENCE

1. Shawyer R. EmDrive Thrust/Load Characteristics. Theory, Experimental Results and a Moon Mission. IAC-19-C4.10.14.

15 The Technology of Hope

In the previous three chapters, we have seen how one 4G engine design can be employed in three different vehicles to provide a wide selection of space and terrestrial transportation solutions. They also provide planetary protection against asteroids and will enable solar power satellites to provide low-cost, renewable energy worldwide. But there will still be an urgent need to combat global warming. It has been proposed that a huge sunshield could be constructed at Lagrange Point 1, between the Sun and the Earth [1]. The enormous number of components for this sunshield could be flown along a simple vertical flight path to this stable point in space using the Heavy Launch Vehicle. However, deep uncertainties have been expressed by a number of experts when solar radiation modification is considered for the whole of the Earth's surface. We should therefore look for a much more modest and controlled solution by reintroducing first-generation EmDrive propulsion. With continuous sunlight, this can provide continuous thrust, which would enable much smaller sunshields to provide controlled cooling for selected areas of the Earth's surface. These sunshields would need to be constructed in medium Earth orbits, typically at an altitude of around 20,000 km. As with the proposed solar power satellite (SPS), the sunshield would utilise robotic construction, which would benefit from the precise in-orbit manoeuvring enabled by the 8 fully gimballed thrusters in the HLV. A nominal 2,000-tonne sunshield of 3.6 km diameter could be constructed as a polygon, comprising 40 lightweight 136-μ Kapton shades, unfurled from 40 lightweight semi-rigid spokes. At the tip of each spoke would be a Space Tug, incorporating a Freon-cooled first-generation EmDrive engine, powered by a typical communications satellite solar array. The engines would provide continuous outward thrust to maintain tension on the spoke. With the engines fully gimballed, the whole sunshield attitude could also be continuously altered to maintain shade over a single point on the Earth's surface for a short period of daylight hours. The engines would also provide the necessary thrust to counter the solar radiation pressure on the sunshield and thus maintain stationkeeping. A total of 41 HLV launches would complete the construction of this sunshield. Note that unlike many geoengineering proposals to provide solar shading to the Earth, the use of EmDrive propulsion allows for total control of the amount of shade required. It also gives the ability to change the orbit to cover another point on the Earth's surface and the ability to remove the sunshield, if it is no longer required, and send it into the Sun, the ultimate waste incinerator.

The EmDrive engine proposed for the Space Tug is based on a 2.45 GHz TE015 cavity to maximise the Q value. Using an active cooling system to minimise the operating temperature range and not requiring the complex modulation of later-generation designs means that a magnetron could be used to provide the microwave power. These simple, lightweight sources have a higher efficiency than the solid state amplifiers (SSPAs) usually used. An experimental cavity shown in Figure 15.1, with

DOI: 10.1201/9781003456759-15

FIGURE 15.1 Experimental 2.45 GHz TE015 cavity (Brian Crighton).

internal diameters of 310 mm and 151 mm and a resonant length of 426 mm, has been built and tested whilst powered by an 800 W magnetron. The E and H field plots are shown in Figure 15.2, and the guide wavelength and wave impedance plots are given in Figure 15.3.

The Space Tug, with body dimensions of 0.8 m × 0.8 m and 2 m long, is shown in Figure 15.4. The thermal radiator has a width of 1.8 m and, when deployed from the base of the tug body, has a length of 6.2 m. A mass budget for the Space Tug is given in Table 15.1.

A 16-panel solar array with a deployed width of 26 m would provide 16 kW of DC power at end-of-life. 14.3 kW would be used by the magnetron to provide 10 kW of microwave power to the thruster. The specified minimum thrust of 5 N from each of the 42 space tugs, would be sufficient to rotate the whole sunshield through 180 degree, within 12 hours. This would allow the amount of shade to be varied from maximum to zero between each daily pass over the required area. An outline diagram of the Space Tug attached to the end of a lightweight aluminium spoke of the sunshield is given in Figure 15.5.

The assembly sequence for the sunshield employs an interesting technique being developed by NASA. Their starshade project [2] makes use of an origami approach to deploy a 36-m diameter telescope shade from a 2.5-m stowage container. A similar method would be used to deploy the hub section for the sunshield, only this time

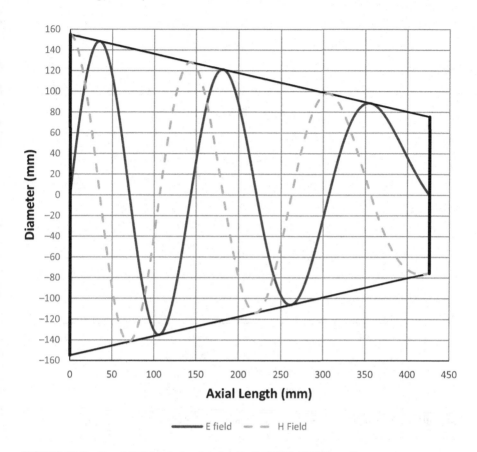

FIGURE 15.2   E and H field phase plots for the 2.45 GHz TE015 cavity.

with a 100-m diameter and using two EmDrive Space Tugs to provide the AOCS. The hub would be carried to orbit as a single HLV payload, and once deployed, it would enable partially unfolded spokes to be docked to the hub. This is illustrated in Figure 15.6.

The 30-m long folded spokes would be partially unfolded by deploying the tug arrays and then using the opposing thrusts of the tug and the HLV that carried it to orbit to unfold the spoke. Once each spoke is partially unfolded, a small tug with robot arms would be used to connect lines from each of the folded shade segments to a reel on the end of the adjacent spoke. When all 40 spokes are in place, the whole sunshield will begin to unfurl, using the spoke tugs to pull out the spokes to their full length of 1.8 km. The radial thrust from each tug would be balanced by an equal and opposite thrust from the tug at 180 degrees. At the same time, the reels at the end of each spoke would begin to reel in the line attached to the corner of the adjacent shade. Thus, a slow, elegant ballet, using the troupe of 40 tugs, would deploy the complete 3.6-km diameter sunshield. A mass budget for the sunshield is given in Table 15.2.

Typical areas in urgent need of overall temperature reduction are major glaciers and cities. For instance, Delhi, in India, recently recorded a record high temperature

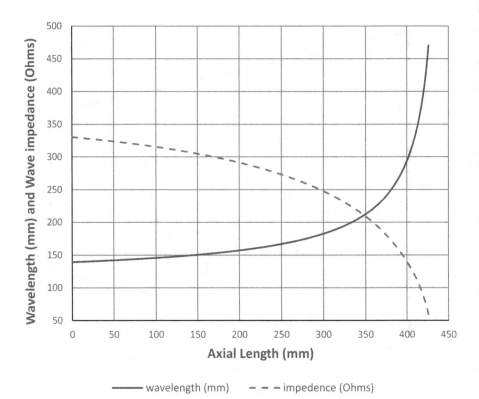

FIGURE 15.3 Guide wavelength and wave impedance for the 2.45 GHz TE015 cavity.

TABLE 15.1

Space Tug Mass Budget

Item	Mass (kg)
Thruster	42
Magnetron	10
Array	547
Structure	50
Pump	20
Radiator	115
Total	784

---

of 47.2°C for this densely populated city. From an altitude of 20,250 km, giving an orbital period of exactly 12 hours, the 2,000-tonne orbital sunshield would produce an eclipse path almost 100 km wide. The penumbral shade would last for 26 minutes, with 1 minute of total eclipse. A similar eclipse would occur at a point on the opposite side of the Earth over the Pacific Ocean. Clearly, a lot of study and consultation

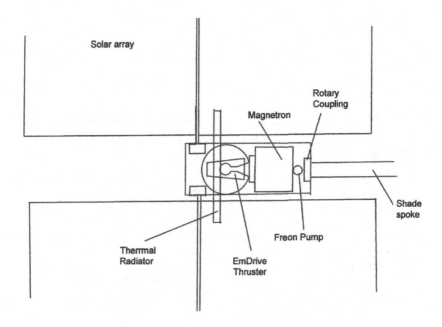

FIGURE 15.4 Space tug.

are needed to optimise the number, size and orbit of sunshields for any particular point on the Earth's surface. However, this simple example shows how relatively small, EmDrive-controlled orbital sunshields could provide significant relief for a city population of around 32 million people.

The type of construction proposed for the sunshield could also be used for a SPS in GEO. Perovskite solar cells can be easily deposited onto flexible surfaces, such as those used in the sunshield. These latest materials now give similar performance to the rigid silicon cells used in present-day solar arrays. An IAC 19 paper from the Korea Aerospace Research Institute described an SPS with a solar array area close to that of the shade area for the 2,000-tonne sunshield [3]. The microwave antenna could also be constructed in a similar manner, with the microwave elements printed onto the flexible substrate. This would form a phased array, enabling fine pointing of the beam, controlled by a pilot signal from the ground station. The 1-km diameter antenna would use 10 spokes and a smaller hub structure. A plan diagram of the SPS is shown in Figure 15.7.

A major problem with any SPS design is the need to keep the solar array pointed towards the Sun while keeping the microwave antenna pointed towards the rectenna on Earth. In a number of designs, this requires a rotating coupling, which carries the total power output of the solar array. This is a challenging requirement, which can be overcome by using the EmDrive tugs to rotate the whole satellite 360 degrees around the hub axis. The microwave antenna is tilted by +/–90 degrees using the tugs at each end of the antenna spokes. This is illustrated in the SPS elevation diagrams in Figure 15.8, for a full 24-hour rotation of the Earth. DC power would be transmitted

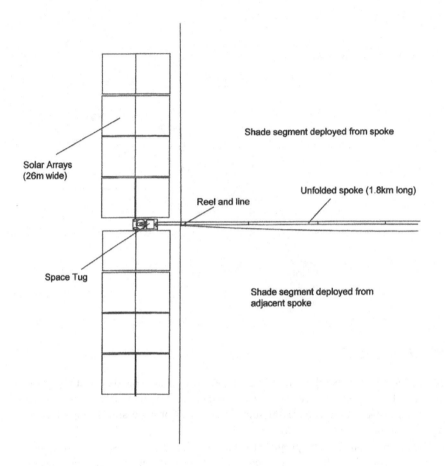

FIGURE 15.5   Space tug attached to the sunshield.

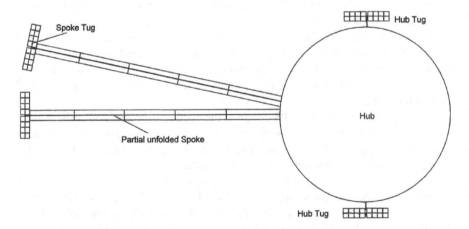

FIGURE 15.6   Deployed sunshield hub with 2 of the 40 partially unfolded spokes.

TABLE 15.2

Sunshield Mass Budget

Item	Mass (T)	No Off	Total (T)
Spokes	1.12	40	45
Tugs	0.748	42	33
Shade segments	48	40	1,920
Hub	50	1	50
Total			2,048

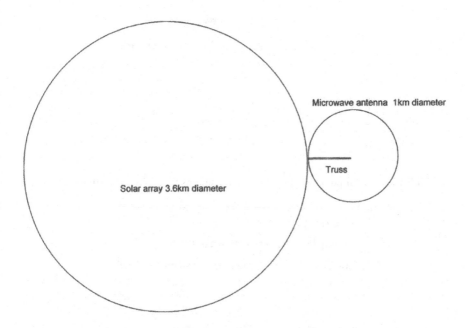

FIGURE 15.7   Plan outline of a solar power satellite

over rigid power lines in the truss structure, linking the solar array to the antenna with flexible braids at the ends to enable tilting of the antenna.

A mass budget for the SPS is given in Table 15.3.

A major function of the space tugs is to counter the Sun's radiation pressure. For the total deployed area of both the solar array and antenna, this amounts to 127 N. The total thrust from the 54 tugs is 270 N, thus providing an ample margin for attitude control.

With the 1-km diameter, 5.8 GHz antenna on the SPS, a power of 2 GW can be beamed to a 4-km diameter rectenna on the ground. The IAC-19 paper suggests the power receiving stations could be installed in the 4-km wide Demilitarised Zone, between North and South Korea. It was stated that up to 120 GW of SPS power would meet the energy requirements for both North and South Korea combined, and

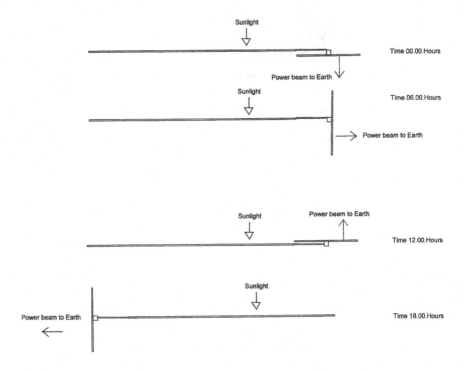

FIGURE 15.8   Solar power satellite elevation diagrams for a 24-hour earth rotation.

TABLE 15.3
Solar Power Satellite Mass Budget

Item	Mass (T)	No off	Total
Array spokes	1.12	40	45
Tugs	0.784	54	42
Array segments	57.03	40	2,281
Array hub	50	1	50
Truss	20	1	20
Antenna segments	15.1	10	151
Antenna hub	14	1	14
Antenna spokes	0.31	10	3.1
Total			2,606

it was implied that the concept could therefore become a much-needed political and economic bridge between the two countries.

The Korean study illustrates perhaps that the most important reason for developing fourth-generation EmDrive is that it would become the enabling technology for a more peaceful and stable world. A less environmentally stressed world should produce a more peaceful world. Indeed, this was one of the reasons put forward by Colonel Coyote Smith of the USAF (see Chapter 7) when he was heading up the SPS study in the Pentagon. By 2008, it had become clear that the consumption of fossil fuels could not continue at its present rate without becoming a future source of conflict. One of the prime responsibilities of any defence and intelligence organisation is to look forward and discern why conflicts may arise and what developments could help to minimise them. The need for solar power from space was pretty obvious, and the need to dramatically reduce launch costs to GEO was therefore paramount. Hence the interest of the US intelligence community in EmDrive.

The three fourth-generation vehicles, described in the preceding chapters, together with the first-generation EmDrive Space Tug, cover most of the applications needed to proceed to a peaceful future. However, human nature always seems to produce conflict, and the benefits of EmDrive may not happen quickly enough. Also, the latest study of the interconnection between climate and agriculture at Cambridge University predicts 6 billion deaths by starvation by 2100 [4]. It is therefore prudent to look towards a future fifth-generation EmDrive to see how it could mitigate the worst catastrophic effects of climate change or conflict. This is where John Prewer has put forward a number of interesting solutions to meet the essential requirements of survival, as originally put forward by his early mentor, Buckminster Fuller. We have seen in Chapter 14 how a Mars base could become self-sufficient on just sunlight and water, and therefore, if necessary, such bases could become early demonstrators for survival technologies on Earth. The need to move whole communities away from areas of catastrophe or to avoid extreme climate-induced weather conditions led to the concept of flying villages. They were termed Arial Refuge Communities (ARCs) and would comprise 1.6-km diameter domes, housing up to 7,000 people. The geodesic dome roof would be used as a rectenna to receive microwave power beamed down from the nearest SPS. The lift for the ARC would be provided by the higher humidity and air temperature inside the dome compared to the external air. The attitude, altitude and position would be controlled by fifth-generation, room-temperature superconducting, EmDrive thrusters, cooled by conventional Freon thermal control systems. An illustration of an ARC and the flying vehicles necessary to provide personal transportation to and from the Earth's surface is given in Figure 15.9.

The homes within the ARC are lightweight, single-storey courtyard designs. The whole community would be self-sufficient in food using the soil-less, aeroponic farming system proposed for the Mars base. Electrical power from the SPS would be plentiful and low-cost. For a redundant power supply source, the triodetic floor structure could be used as a rectenna for secondary power transmission from deep borehole geothermal power stations. An internal view of the ARC is given in Figure 15.10.

The room-temperature superconducting materials that would be necessary for coating the internal walls of the cavities of the fifth-generation EmDrive thrusters are currently the subject of a number of research programmes. It is predicted that such materials, with the performance of Yttrium barium copper oxide at 20°K, but operating at room temperature, will become available within the timescale we are now considering. Similarly, new battery technology for the personal air vehicles shown in Figure 15.11 will also be developed. The current prediction is a specific energy of 1,200 Whr/kg for the lithium-air batteries, which are presently being researched by the Argonne National Laboratory and the Illinois Institute of Technology in the US [5]. This would provide total flight times, without recharging, of many hours.

FIGURE 15.9 Arial Refuge Communities and personal air vehicles (Alan Gilliland).

FIGURE 15.10 Internal view of Arial Refuge Communities (Alan Gilliland).

With large tanks of LH_2 no longer needed to cool the thruster cavities, the design of such vehicles can be quite radical. It is envisaged that the lightweight airframe would be manufactured from high-tech glass, which is currently under development. Operation would be by voice control, and with four fixed thrusters powered by SSPAs, there would be no moving parts. Thus, together with the advanced batteries and new materials, operational lifetimes could be targeted at centuries. We would be moving into a long-term recyclable future with less wasteful manufacturing.

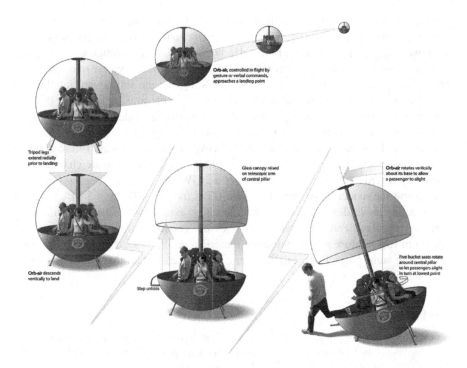

Orb-air, controlled in flight by gesture or verbal commands, approaches a landing point

Tripod legs extend radially prior to landing

Glass canopy raised on telescopic arm of central pillar

Orb-air rotates vertically about its base to allow a passenger to alight

Five bucket seats rotate around central pillar to let passengers alight in turn at lowest point

Orb-air descends vertically to land

Step unfolds

FIGURE 15.11 Fifth-generation personal air vehicle (Alan Gilliland).

This book has told the EmDrive story from its initial concept as a possible solution to a serious problem with the UK nuclear warhead, under development in the 1970s, through the development of a basic theory of operation using established Physics and well-understood principles of microwave engineering. As always in such research and development projects, the theory was then extensively tested using experimental and demonstration thrusters. This work was funded and monitored by the UK government. Once the work came into the public domain, a lot of controversy arose, particularly when other experimenters tried to replicate the results without understanding the theory or realising the considerable engineering challenges posed in designing, building, tuning and testing the thrusters. The UK government granted export licences to the US and Israel, and reports of Russian and Chinese projects were received. A flight thruster programme was completed, and the results were delivered to Boeing under a Technology Assistance Agreement set up by the US State Department and covered by a contract between SPR Ltd. and Boeing's Phantom Works in Huntington Beach. The considerable financial and technical resources required to take the technology beyond the first-generation thrusters and the experimental second-generation superconducting thrusters were beyond SPR capabilities. We therefore continued the work on a theoretical basis and completed a number of design studies for air, space and marine transport applications. These studies showed the need to solve the difficult problem of accelerating high-Q cavities without losing thrust and the unexpected problems caused by the need to load a thruster on start-up.

A third-generation thruster design was invented and patented, and the theoretical basis was subjected to limited experimental work that confirmed predictions. This work was presented to UK government defence organisations through a series of lectures at the UK Defence Academy at Shrivenham. Finally, the release of unidentified arial phenomena (UAP) sightings by the US Navy and the subsequent intelligence report spurred on the theoretical work and resulted in the fourth-generation thruster design. A fourth-generation, liquid hydrogen-cooled, engine design was then incorporated into three vehicle design studies. The Deep Space Probe, Heavy Launch Vehicle and Personal Spaceplane seemed to cover all the immediate future applications. These included enabling solutions to global energy and climate change problems. In particular, the reduction in mass of a 2 GW SPS from 6,700 tonnes to 2,600 tonnes and the reduction in launch costs to GEO from $1,000/kg to $8.4/kg offered encouraging methods to solve these problems. However, as with all engineering projects, there is always an improvement to be made. The possibilities offered by potential room-temperature superconducting materials and advanced battery technology then led to some projections into the future and the realisation that EmDrive truly is the technology of hope.

REFERENCES

1. Baum C.M et al. Between the Sun and us: expert perceptions on the innovation, policy and deep uncertainties of space based solar geoengineering. Renewable and sustainable Energy Reviews. 158(2022)112179. Elsevier. Feb 2022.
2. NASA Space Origami. go.nasa.gov/2m1QT6B
3. Joon-Min Choia. Guee Won Moonb. Conceptual Design of Korean Space Solar Power Satellite. Korea Aerospace Research Institute. IAC-19-C3.1.2.
4. Richards C E et al. *International risk of food insecurity and mass mortality in a runaway global warming scenario*. Department of Engineering, University of Cambridge UK. 150 June 2023 103173.
5. Kondori A et al. *A room temperature rechargeable Li_2O-based lithium-air battery enabled by a solid electrolyte*. Illinois Institute of Technology, Chicago and Argonne National Laboratory, Lemont, IL. SCIENCE Vol 379(6631) 2 Feb 2023.

Index

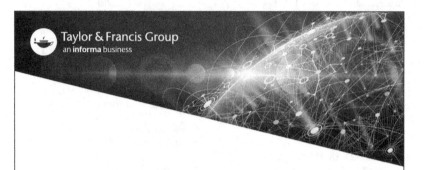

Printed in the United States
by Baker & Taylor Publisher Services